Managing
Innovation

Pergamon Titles of Related Interest

Dewar INDUSTRY VITALIZATION: Toward A National
Industrial Policy
Fusfeld/Haklisch INDUSTRIAL PRODUCTIVITY AND
INTERNATIONAL TECHNICAL COOPERATION
Fusfeld/Langlois UNDERSTANDING R&D PRODUCTIVITY
Hill/Utterback TECHNICAL INNOVATION FOR A
DYNAMIC ECONOMY
Nelson GOVERNMENT AND TECHNICAL PROGRESS:
A Cross-Industry Analysis

Related Journals*

BULLETIN OF SCIENCE, TECHNOLOGY & SOCIETY
COMPUTERS AND INDUSTRIAL ENGINEERING
COMPUTERS AND OPERATIONS RESEARCH
SOCIO-ECONOMIC PLANNING SCIENCES
TECHNOLOGY IN SOCIETY
WORK IN AMERICA INSTITUTE STUDIES IN PRODUCTIVITY

*Free specimen copies available upon request

Managing Innovation

THE SOCIAL DIMENSIONS OF CREATIVITY, INVENTION AND TECHNOLOGY

EDITED BY

SVEN B. LUNDSTEDT
E. WILLIAM COLGLAZIER, JR.

FOREWORD BY

FRANK PRESS

PUBLISHED WITH THE ASPEN INSTITUTE FOR HUMANISTIC
STUDIES AND THE OHIO STATE UNIVERSITY

PERGAMON PRESS
New York   Oxford   Toronto   Sydney   Paris   Frankfurt

Pergamon Press Offices:

U.S.A.   Pergamon Press Inc., Maxwell House, Fairview Park,
                    Elmsford, New York 10523, U.S.A.

U.K.   Pergamon Press Ltd., Headington Hill Hall,
                    Oxford OX3 0BW, England

CANADA   Pergamon Press Canada Ltd., Suite 104, 150 Consumers Road,
                    Willowdale, Ontario M2J 1P9, Canada

AUSTRALIA   Pergamon Press (Aust.) Pty. Ltd., P.O. Box 544,
                    Potts Point, NSW 2011, Australia

FRANCE   Pergamon Press SARL, 24 rue des Ecoles,
                    75240 Paris, Cedex 05, France

FEDERAL REPUBLIC Pergamon Press GmbH, Hammerweg 6
OF GERMANY   6242 Kronberg/Taunus, Federal Republic of Germany

Library of Congress Cataloging in Publication Data

Main entry under title:

Managing innovation.

    "An Aspen Institute book."
    Includes index.
    1. Technological innovations--United States--
Addresses, essays, lectures. I. Lundstedt,
Sven B., 1926- . II. Colglazier, E. William
(Elmer William)
HC110.T4M275 1982   658.4'06   82.293
ISBN 0-08-028815-4   AACR2

Printed in the United States of America

Unless there is a speeding up of social invention or a slowing down of mechanical invention, grave maladjustments are certain to result —*From Recent Social Trends in the United States, Report of the President's Research Committee on Social Trends, 1932.*

Not everything new, of course, comes from mechanical invention. There are social inventions also, as, for instance, proportional representation, social insurance, the holding company, and the League of Nations. Some social changes originate then from social inventions. Social inventions may have been precipitated by mechanical inventions, as, for instance, the Interstate Commerce Commission was caused by the railroads. The connection between the social invention and the mechanical is not so close in the case of the juvenile court, which results from changes in the urban family due in turn to the mechanical forces that produced city life. Some social inventions are so far removed from mechanical inventions that any connection is scarcely discernible. Such would be the case with the invention of the parole of prisoners. On the other hand, some social inventions cause mechanical inventions. Thus, a sales tax may bring out a new token money. A zoning law, as in New York, may force architectural devices to be used to modify the skyline, or an antinoise campaign may cause the invention of a rubber horseshoe used in connection with milk wagons on early-morning deliveries. . . . Indeed, the more one studies the relationship between mechanical and social inventions, the more interrelated they seem — *William Fielding Ogburn, Journal of Business of the University of Chicago IX, no. 1 (January 1936).*

For innovation means the creation of new value and new satisfaction for the customer. Organizations therefore measure innovations not by their scientific or technological importance, but by what they contribute to market and customer. They consider 'social innovation' as important as 'technological innovation.' Installment selling may have had greater impact on economies and markets than most of the great 'advances in technology' in this century —*Peter F. Drucker, Wall Street Journal, Friday, February 6, 1982.*

CONTENTS

Foreword
  Frank Press   ix

Preface
  Sven B. Lundstedt   xi

Introduction
  Sven B. Lundstedt and E. William Colglazier, Jr.   xiii

Chapter
  1 Social and Technological Innovation
    Harvey Brooks   1

  2 Conservative and Radical Technologies
    Thomas P. Hughes   31

  3 Innovation and the Grants Economy
    Kenneth E. Boulding   45

  4 The Innovative Milieu
    Ralph Landau   53

  5 The Human Side of Technological Innovation: Labor's View
    Iris J. Lav and Stanley H. Ruttenberg   93

  6 Government and Innovation
    Daniel De Simone   111

  7 Politics, Politicians and the Future of Technology
    Clarence J. Brown   123

  8 Environmental Regulations and Technological Innovation
    Robert D. Hamrin   148

  9 The National Aeronautics and Space Administration: Its
    Social Genesis, Development and Impact
    Karl G. Harr, Jr. and Virginia C. Lopez   170

10 Absorption and Adaptation: Japanese Inventiveness in
   Technological Development
   Tetsunori Koizumi   190

11 Human Factors Affecting Innovation and Productivity
   Michael Maccoby   207

12 The Jamestown Experience: A Case Study in Labor-
   Management Cooperation
   Stan Lundine   215

13 Social Invention and Innovation as Partners in Technological
   Progress
   Thomas H. Moss   229

14 On Using Intellectual Resources
   Sven B. Lundstedt   242

Index   247

About the Authors   250

FOREWORD

After some twenty years of talk but no action, the nation seems
ready to address the issue of technological innovation. The public
is now concerned about U.S. technological lag, perhaps due to the competitive prices, high quality, and advanced design of imports in consumer electronics, integrated circuits, automobiles, cameras, airplanes,
and other middle and high technology products. Labor is concerned
about growing unemployment in low and middle technology industries,
and is increasingly critical of management's lack of investment in modern equipment and production lines. Industry, in addition to its criticism
of government economic and regulatory policy as impediments to innovation, is now examining the impact of its own policies, particularly
its management reward systems and their effect on innovation. Perhaps
most important of all, Congress and the Executive Branch are more
worried than ever about productivity and inflation and seem ready to
address the government's significant impact on these matters. It is quite
likely that regulatory, tax, patent, antitrust, trade, and R&D policy will
see major changes, motivated primarily by concern over the related
issues of innovation, productivity, inflation, unemployment, and balance of trade. The nation seems poised to act on these matters and
benign neglect will be replaced by a debate on the question of what we
should do.

Should we disinvest in noncompetitive industries and transfer resources to the high technology areas? What about retraining and providing in other ways for dislocated workers? Should tax policy generate
capital growth and spur investment generally or should tax benefits be
targeted towards innovation and productivity improvements? To what
extent should economic impact be a predominant factor in analyzing
proposed regulations, and can we attach confidence to and use risk
analysis methodologies? Should we initiate incentives and policies to
spur exports? Can we develop new institutional relationships between
government, universities, and industry to enhance innovation? Can we

help developing countries with their modernization plans without placing our own industries at risk in the years ahead? These are among the
issues we will be debating in the years ahead as we enter a second
industrial revolution, one based on new technologies and new attitudes
towards innovation and industrial policy.

The competition between the industrial democracies in the newly
formed high technology industries is likely to be fierce. Trade barriers,
impediments limiting scientific communication and information flow,
and pressures on multinational corporations may be among the negative consequences of the new high technology era. We hope these issues
can be addressed before they create economic havoc and political dis-
unity. On the other hand, new innovations such as production sharing,
in which various segments of the total product are each in a different
country which can handle its segment most economically, may help
reduce protectionist pressures.

Few doubt that the United States cannot perform superiorly in science
and technology—the beginning of the innovative process. By almost any
criterion, the United States remains the world leader in scientific progress; almost every important scientific advance with commercial potential has been developed here. The steps that follow R&D in the development of new products and processes—steps that depend on the
availability of capital, wise investment decisions, and benevolent government policy—will now be addressed by the nation, and an industrial
policy fostering innovation and investment will emerge.

Now that there appears to be a national readiness to do the right
things to foster innovation, this volume is particularly timely. The collection of papers includes contributions by historians, economists, entrepreneurs, government officials, and scientists and engineers—each
author viewing the innovation process from a different vantage point.
An important feature is the beginning discussion of the relationship
between the social aspects of innovation, (e.g., the organizational, management, and human aspects) and artifact invention and innovation.
Although anecdotes abound involving technological advances, which
because of social decisions have led to commercial successes or failures,
the entire process of innovation with adequate recognition of the role
of social invention has been insufficiently studied. In broadening the
conceptual basis of technological innovation to include social aspects,
this volume makes a special contribution to the bountiful literature on
the subject.

Frank Press

PREFACE

The concept upon which this book is based recurred to me during the summer of 1979 while in residence at The Aspen Institute for Humanistic Studies in Aspen, Colorado. Technological innovation and its relationship to industrial productivity and economic growth has always been an important subject. But an aspect of it seemed to be underemphasized in the mainstream of that line of thought. Sufficient attention has not been given to social invention and innovation. How people respond creatively to artifact invention and innovation with social invention and innovation is also of critical importance.

Social inventions and innovations are ubiquitous. Yet, because they are often obscured by other names, or taken for granted, their value is not always recognized. We do not have to look far for typical examples. The modern corporation is a particularly successful social invention and innovation. It is one of the foremost reasons why technology has been so effectively exploited in this century. Technological, industrial and economic development have also brought on social inventions and innovations which have served to reduce the unwanted side effects of such development. Trade unions and government regulation are examples. The Taft-Hartley Law is a particularly interesting social invention and innovation for managing industrial conflict. The National Environmental Policy Act was "invented" to limit environmental damage coming from industrial productivity. Much legislation is social invention and innovation; the distinction between invention and innovation will become clear later in this book.

This collection of original papers calls attention to the synergism which exists between these different kinds of invention and innovation and their relation to industrial development and economic growth. A basic thesis of this book is that such growth will be more successful when there is a creative balance between these different forms of invention and innovation.

A grant from The Atlantic Richfield Foundation has enabled this book to be prepared, and enabled a developmental conference to be held which served initially as a starting point for this enlarged volume of papers. The book has become a separate project apart from the conference so it is not in any sense merely a conference proceedings. The original conference was held November 7–9, 1980 at The Aspen Institute Wye Plantation campus on the Eastern shore of Maryland in cooperation with the Institute's Program in Science, Technology and Humanism.

We wish to recognize and to thank the following distinguished conference participants who, in addition to several of the chapter authors, took part in the discussion: Charles Cook, Senior Vice President SRI International; William Cunningham, Economist, AFL-CIO: Paul Doty, Senior Fellow, Aspen Institute and Professor and Director of The Center for Science and International Affairs, Harvard University; Amitai Etzioni, University Professor, George Washington University; N. Bruce Hannay, Vice President, Bell Telephone Laboratories; Paul Horwitz, Senior Scientist, Bolt, Beranek and Newman, Franklin Lindsay, Chairman, ITEK Corporation; Jessica Tuckman Mathews, Editorial Staff, The Washington Post; Frederick J. Milford, Associate Director for Research, Battelle Columbus Laboratories; Howard A. Slack, Vice President for Technology, The Atlantic Richfield Company; Ezra Vogel, Professor of Sociology, Harvard University; Charles Zraket, Executive Vice President, Mitre Corporation; Abe Spinak, Assistant Director, National Aeronautics and Space Administration; Dorothy Zinberg, Special Adviser, Aspen Institute and lecturer, Harvard University (affiliations are as of the time of the seminar).

We wish to recognize the kind support and generosity of The Atlantic Richfield Foundation for its financial support and for the additional support provided by The Ohio State University and its School of Public Administration and Research Foundation. I wish to also thank Mr. Joseph Slater, President of the Aspen Institute, for his encouragement. And we acknowledge the distinguished endorsers whose names appear on the jacket of this book.

Paul Doty, Director of the Program in Science, Technology and Humanism and William Colglazier, then Associate Director of the program, were my venture partners. I wish to thank Harvey Brooks, William Colglazier, Paul Doty and Paul Horwitz for serving on the planning committee for the conference. I especially want to thank Fred A. Nelson of The Atlantic Richfield Foundation for his helpful cooperation. I gratefully acknowledge the contributions of William Colglazier as co-editor and conference rapportuer, and of Harvey Brooks and Paul Doty who shared with me the task of moderating the conference. And finally I wish to thank my wife, Jean S. Lundstedt, for her kind assistance.

Sven B. Lundstedt

INTRODUCTION

Sven B. Lundstedt and
E. William Colglazier, Jr.

Technological innovation is the process that begins with an inventor's insight and ends with a new product or technique in the marketplace.[1,2] Many scholars believe that technological change and new knowledge are more important contributors to economic growth over the long run than the classical variables of land, labor, and capital. While the variance is large, the average rate of return on private investment in technological innovation has been estimated to be twice that of capital. Even more significant, the return to society has been estimated to be double that accruing to the private investor.[3]

Being the world leader in technological innovation has produced enormous material benefits for the United States in the twentieth century. Yet, as expressed in a recent *Newsweek* article entitled "Innovation: Has American Lost Its Edge," "technocrats have begun to fear that America's lead in innovation—and the vaunted U.S. technological superiority that it spawned—may be withering."[4] Although innovation is sufficiently idiosyncratic that it should be examined on an industry by industry basis, some indicators of a general U.S. decline over the past decade are lower investment in research and development as a percentage of Gross National Product, a declining share of the world technology market, a smaller share of patents going to U.S. citizens, and a productivity growth rate and capital investment rate lower than that of most industrial nations.[5] Also worrisome is the perception that current American management practice is weighted toward the financial rather than the entrepreneurial, leading to a focus on near-term goals and an increased aversion to risk taking.[6]

The Carter Administration, as its four predecessors, initiated major studies on industrial innovation but in the end recommended rather limited policies to try to alter the situation. To the disappointment of some of his advisors, President Carter's message on innovation did not even mention tax and economic policies. Rather than studying the problem further, the Reagan Administration responded to the U.S. economic malaise with its tight monetary policies and its version of "supply-side" macroeconomic policies. The result was tax cuts for consumers as well as business and significant reductions in non-military government expenditures, including research and development. One Republican critic stated, "You can't try to stimulate investment and research and development through tax breaks (which are themselves desirable) and then vitiate the beneficial effects by huge consumer tax cuts that loosen fiscal policy and drive interest rates sky high."[7] Although there obviously has not been societal consensus on the appropriateness of policies attempted over the past few years, politicians have generally recognized that revitalizing economic growth through the fostering of technological innovation is a critical imperative facing the U.S. in the 1980s.

The purpose of this book is to explore new directions that support for industrial innovation can take through the next decade in the United States. The special focus is on the synergism between social and technological innovation, or stated more directly, how new arrangements in social structures or processes can facilitate the use and diffusion of new technological products or processes and thereby generate increased productivity and economic growth.

Social invention and innovation consist of changes in the communication, organizational, and power relationships among people in various institutions and society. They can occur on any scale, from new social relations or activities in small groups to new organizations or social structures on a global scale. They are the natural and ubiquitous companions of technological invention and innovation, but are not widely understood or even recognized as a rule. For a variety of reasons, a purely technological bias overshadows the effects of psychological, social, and cultural forces that surround and often determine the form and direction of invention and innovation. As stated by Harvey Brooks of Harvard University,

> It has been traditional to define technology in terms of its physical embodiments, as novel physical objects created by man to fulfill certain human purposes. . . . this is too limited a view and one that is becoming increasingly obsolete. . . . technology must be sociotechnical rather than technical, and a technology must include the managerial and social supporting systems necessary to apply it on a significant scale.[8]

Brooks has identified two goals that can be served by technological and social innovation in addition to improving material living standards and economic productivity. One is to improve the quality of life as related to the natural evironment, social relationships and a more egalitarian society. The other is to improve the quality of working life as related to greater self-development and self-realization in work and the reduction of authoritarian relationships. The potential conflict between these three goals can be substantially reduced by combining social innovation with technological innovation, which may be especially needed when the social, political, and economic environment is rapidly changing.

Perspectives from the Wye Conference

The conference held at Wye, Maryland, in November 1980, was not designed to reach consensus or make recommendations, but rather to allow thirty participants from government, industry, labor, the media, and universities to explore the connection between social and technological innovation. Although discussion frequently returned to the macro-economic policies of government, additional themes did emerge. A brief recapitulation serves to introduce the following chapters where these ideas are elucidated in detail.

The participants discussed the difference between conservative and radical technological innovations and the relative importance of both to economic growth. Conservative innovations, as defined by Thomas Hughes of the University of Pennsylvania, interact supportively with current institutions designed to nurture technology. Radical technological innovations, on the other hand, cause sharp changes, disruptions, and displacements in existing organizations, and they tend to come from the lone inventor or entrepreneur. While radical innovations may be critically important from the historian's perspective, the steady stream of conservative innovations that can come from creative large companies can add up to significant economic growth. The strength of the U.S. system may lie in its ability to reap the benefits of both.

A recurrent view among many participants was that the supply side of the economy needs attention, that is, the United States has been overconsuming and underinvesting for too long. The highest priority social objective for the 1980s should be "increasing the pie," making questions of how to slice it secondary. The problem was not seen to be a shortage of research and development or new ideas, but the lack of incentives and too many uncertainties further downstream in the innovation process. Tax and economic policies, many felt, should be the major tools for increasing the rate of technological innovation. In other words, wealth creation is the primary way to foster innovation, and that requires increased incentives for the entrepreneur (a lower capital gains tax and an accelerated depreciation allowance, as was included in the

1981 tax cut legislation) and reduced governmental uncertainties (stable economic and regulatory policies). The fact that venture capital availability surged following the capital gains tax reduction in 1978 was cited as proof of the potential impact of tax reductions on innovation. In addition, some participants favored a more rigorous cost-benefit analyses being applied to government regulations.

Another theme espoused by many participants was the importance of increasing communication and trust between management and labor and joint grass-roots planning (upward flowing and micro-social innovation rather than central control and macro-social invention). Labor shares the desire for economic growth and recognizes the key role of technological innovation. But labor is more sensitive to the fact that innovation may disrupt the lives of individual workers directly and indirectly. Therefore, many participants stressed that companies and government must help cushion workers from the impact of changes in technology and changes in comparative advantage between industries. Also, the economic benefits of productivity increases should be shared between profits, wages and prices. Regarding regulations, it was stressed by some that the cost of repairing environmental damage is often greater than the cost of avoiding it. What is needed, many participants believed, is ideological disarmament between management and labor. Increased management attention to human concerns and to consultative management (which is not the same as co-determination) has proven to be a source of productivity increases and improved product quality. While not being very strong in conceiving technological innovations, the Japanese have achieved rapid economic growth through social and managerial innovations and through adaptation of new technological ideas from foreign sources. The United States also needs to be more creative managerially and to be more open to innovations originating overseas.

A feeling among nearly all of the participants was that there is an emerging national consensus on the need for increased investment and productivity. This consensus, however, is a fragile one that must be fertilized by greater management attention to human concerns, communications, and grass-roots planning in the work place.

Overview of the Chapters

Frank Press, the former Science Advisor to President Carter and now President of the National Academy of Sciences, has already ably introduced in the Foreword the multitude of policy issues surrounding the strengthening of the innovative capacity of U.S. industry and its international competitiveness. He was a key participant during the past four years in efforts to comprehensively review and raise these critical issues to the U.S. political leadership.

The first chapter by Harvey Brooks of Harvard University provides an integrative assessment of the social dimensions of invention and innovation, the central theme of this book. With numerous examples, he characterizes innovations according to the importance of their social dimensions, and then classifies social inventions and innovations as being either market, managerial, political or institutional. Innovative marketing and delivery systems, for example, aid new technologies in penetrating the market and existing technologies in expanding their market share. Managerial inventions and innovations change work organization to improve labor productivity, the quality of a product or service, and the quality of working life. (These changes are mutually reinforcing, not conflicting alternatives.) He suggests that the Japanese have excelled in managerial innovations, especially in constructive use of worker participation. Political innovations often take the form of legislation aimed at achieving some social objective by making the private incentives more congruent with public benefit. He suggests that the greatest need for political innovation lies in improving the capacity of the political system to integrate conflicting interests into a sustainable program of action. Institutional innovation in his classification refers primarily to changes at the working level of individuals and voluntary groups, where a fruitful area for small-scale experimentation is developing grass-roots innovations in consensus building.

Thomas Hughes of the University of Pennsylvania discusses in Chapter 2 his distinction between conservative and radical technologies. He suggests that American invention and technological development are now generally conservative in effect, producing a great deal of inertia that has led to lagging productivity. As he states, "Technologies and institutions tend to look alike after a while, that is, it is difficult to change one without the other." He asks if re-industrialization can be achieved without radical technological innovations, those that are disrupting to the status quo, and then seeks mechanisms to relieve the anxieties that foment implacable resistance to this type of change. These mechanisms include cushions to soften hardships of change, technological bridges that facilitate the transition from the old to the new, and mergers that give established institutions a stake in radical transformation.

Kenneth Boulding of the University of Colorado in Chapter 3 probes the grants economy, which is the total system of public and private *one-way* transfers of economic goods that redistribute net worth in society. He sees the major motivations for grants as love and fear. Innovators make grants to society by absorbing risks, and they are motivated by seeking self-esteem and wishing to benefit humanity as well as gambling for financial reward. A society that does not sustain the benevolent motivations, he concludes, may be more destructive to creative innovation than faulty financial incentives or inadequate public grants for innovators.

Ralph Landau, Chairman of the Halcon Corporation, describes in Chapter 4 the lessons learned from personal experience as an extremely successful technological entrepreneur. It is the entrepreneurial side of innovation that bears the sustantial risk to carry an invention into the market place, and it is this element in the United States that is currently faltering in his opinion. Government policies promoting redistribution of wealth and pervasive regulation have created barriers to innovation. To unleash the entrepreneurial spirit, leading to sustained economic growth, requires in his (and President Reagan's) view relatively free markets, a steady course by government, and tax policies that improve the profit-making environment for the risk–taker. He states that "we must . . . concern ourselves (now) with enlarging the national 'pie', not in how equally we slice up a static or shrinking pie."

Iris Lav and Stanley Ruttenberg explain in Chapter 5 labor's view of the human side of technological innovation. Labor supports innovation and the resultant economic growth that can increase material living standards and provide new employment. Labor is even willing to invest its deferred compensation for the purpose of increasing industrial productivity, but it wants from management and society an equitable sharing of the economic benefits of innovation with the workers implementing the changes and a guarantee that the needs of workers whose lives are disrupted will be met. Displaced workers need adequate compensation and opportunities for retraining. In addition, communication, goodwill, and cooperation between labor and management will facilitate the smooth acceptance of change as well as stimulate an upward flow of innovative ideas from the shop floor.

Daniel De Simone, President of the Innovation Group, has participated in several governmental and private studies of industrial innovation policy. In Chapter 6 he assesses the impact of government on the innovation climate and concludes that the dominant effect comes from indirect policies. These are "tax rates, credits, and depreciation allowances that influence investment and risk taking; monetary policies that affect the money supply and the availability of venture capital; the indiscriminate enforcement of the anti-trust laws on the one hand and the rescue of comatose corporations on the other; and the burgeoning accumulation of regulations that impinge upon every step of the innovation process." Similar to the recommendations of Chapter 4, he favors reduced government intervention in the market and greater incentives in the tax structure for risk takers.

Congressman Clarence Brown of the Joint Economic Committee examines in Chapter 7 the role of government action in shaping or responding to both technological and social invention. The Congressional response to an emerging issue often comes down to whether the government should attempt to legislate and regulate a solution or leave the solution up to the private sector. But because politics deals primarily with short-term pressures, the second and third order effects are typically ignored and usually cannot be foreseen. He therefore cautions political leaders to ask whether their legislative solutions "are sensitive and responsive enough to be capable of linking social innovation and technological progress."

Robert Hamrin of the Senate Budget Committee contends in Chapter 8 that environmental, health, and safety regulations can be designed to stimulate rather than retard innovation: "When regulation is steady and gradual and firms have sufficient time to comply, effective and innovative technical solutions appear." He examines a number of cases where innovations stimulated by firms seeking to comply with environmental regulations have resulted in significant economic and energy benefits. The most beneficial efforts are those directed towards elimination of pollution sources, such as with resource conserving technologies, rather than merely trying to control pollution at the end of production. Hamrin concludes that regulations encouraging innovation in compliance and abatement technologies can achieve their environmental goals in the most cost-effective manner for industry and society.

Karl Harr, Jr. and Virginia Lopez of the Aerospace Industries Association make a case in Chapter 9 that "the greatest innovation in the Apollo program was not the hardware, but the managerial system"[9] of the National Aeronautics and Space Administration (NASA). Although strong public and political support in the 1960s were essential for mobilizing the technological resources to put a man on the moon, the real keys to NASA's success were its innovative management techniques and its systems approach. In directing a network of federal laboratories, industrial laboratories, and university scientists, NASA had to manage many detailed and complex projects in synchronism while ensuring that performance and reliability requirements were met on schedule and that all elements moved in an integrated manner toward the final goal. The NASA management system was an appropriate social innovation for directing a highly complex engineering feat.

Tetsunori Koizumi of the Ohio State University addresses in Chapter 10 the historical reasons for Japan's success in technological development. He finds that it depends on the ways in which the efforts of a population are socially mobilized, which in turn depends on traits deeply rooted in a society's culture. Japanese inventiveness has in part come about through its long tradition and skill at creative assimilation, that is, its ability to absorb and adapt. The legacy of the culture on the modern Japanese character has also led to the emphasis on quality control, craftmanship, communication as a means of consensus-building, and innovations percolating from the bottom up.

Michael Maccoby, Director of the Project on Technology, Work and Character, presents in Chapter 11 the considerable empirical evidence to support the thesis that adequately trained and motivated factory and office workers can contribute to a continual process of innovation. He finds that what is needed for technology to fulfill its promise of productivity is new attitudes and practices by management and by union leadership. Consultative and participative methods of management that encourage grass-roots planning and initiative have been shown to increase productivity, quality control, and job satisfaction among workers in comparison to autocratic, policing styles of management. Those companies at the forefront of managerial innovation include both firms which guarantee job security and emphasize respect for the individual, and large unionized companies which can work cooperatively with a progressive union. One of the key ingredients in all these cases, he concludes, is the high quality of the leadership, one that is motivated by a concern for people.

Congressman Stan Lundine describes in Chapter 12 a pioneering prototype experiment in labor management cooperation that he instigated as mayor of Jamestown, New York. The town in 1970 was suffering from high unemployment, factory closings, and poor labor-management relations. With local government serving as the catalyst, a joint labor-management committee was created to focus on improving the productivity and quality of work life. Similar committees were created in individual plants. By agreeing that productivity gains would be shared between labor and management and that workers would not be laid off following productivity improvements, the joint committee built a participative process that enabled labor and management to solve problems together outside of collective bargaining. The resulting improvements included lower unemployment, better labor relations, increased productivity and a reinvigorated local economy. The success at Jamestown is testimony to the importance of the human factor in improving productivity as well as to the wide applicability of quality of working life (QWL) programs.

Thomas Moss, Staff Director of the House Subcommittee on Science, Research and Technology, stresses in Chapter 13 the need to encourage social invention in order to provide a receptive climate for technological innovation. He sees two areas sorely in need of social inventiveness. The first is mechanisms for dislocation mitigation for workers that are adversely affected. This addresses a fundamental weakness in our present ability to assimilate technological changes. Imaginative solutions are possible, he feels, with cooperative and participative mechanisms of joint labor-management problem solving. The other area is improving our systems to plan for the future, which he believes could be the single most important investment for expanding society's capacity to capitalize on the opportunities for technological innovation. In order to avoid the inflexible and dogmatic character often associated with planning, he recommends broad participatory planning processes at various levels that incorporate flexible feedback mechanisms and facilitate societal communication of goals and intentions.

The last chapter addresses the point of view and contributions from the social and behavioral sciences to the issues discussed in this book. Some social scientists would define social invention as a "new and apparently promising strategy designed to solve some persistent and serious human problem."[10] Some of the most fundamental social inventions and innovations catalogued by social scientists in the twentieth century have had a direct as well as indirect bearing on technological development, invention, and innovation. What is now needed from the social scientists is greater application of ideologically-neutral and open inquiry on the problems of stimulating technological innovation.

We are grateful to the authors for their clear exposition of the pervasive interconnection of social and technological innovation. In presenting these diverse but not incompatible perspectives, we hope to foster wider recognition of the importance of social innovation as one key element in increasing economic productivity and improving the quality of working life in the United States.

Notes

1. A useful operational definition for innovation is:

> A technical innovation is a complex activity which proceeds from the conception of a new idea (as a means of solving a problem) to a solution of the problem, and then to the actual utilization of a new item of economic or social value. Innovation should be distinguished from scientific discovery, which involves the observation of a previously unknown or unobserved phenomenon or the acquisition of new knowledge; although relevant discoveries may be incorporated into the innovation. Innovation should also be distinguished from invention, which is the creation of a novel product or process, or a concept of a means of satisfying a need. The invention, however, may provide the initial concept leading to the innovation. Finally, innovation must be differentiated from diffusion of technology, which one author has defined as 'the evolutionary process of replacement of an old technology by a newer one for . . . accomplishing similar objectives',

but which we have broadened to include the extension, improvement, and wider use of existing technology . . . the period of innovation is assumed to extend over a bounded interval of time, extending from first conception of the idea for the innovation to first realization, when the first commercially successful embodiment of the innovation entered the marketplace.

This quote comes from Battelle Columbus Laboratories, *Interactions of Science and Technology in the Innovative Process: Some Case Studies,* final report to the National Science Foundation, Battelle Columbus Laboratories, Columbus, Ohio, March 1973. Part of the definition comes from S. Meyers and D. G. Marquis, *Successful Industrial Innovations,* NSF 69–17; Washington, D.C.: U.S. Government Printing Office, 1969.

2. Some additional useful references on innovation, representing only a small sample of a large literature, are:

De Simone, D., *Technological Innovation: Its Environoment and Management,* Report of the (Charpie) Panel on Invention and Innovation, U.S. Government Printing Office, 1967.

Kelly, P. and Kranzberg, M. eds. *Technological Innovation: A Critical Review of Current Knowledge.* San Francisco: San Francisco Press, 1978.

Mansfield, E. *et al., The Production and Application of New Technology.* New York: W. W. Norton, 1977.

_____ . *Research and Innovation in the Modern Corporation.* New York: W. W. Norton, 1977.

Mansfield, E. *Industrial Research and Technological Innovation.* New York: W. W. Norton, 1968.

Committee on Economic Development. *Stimulating Technological Progress.* New York: CED, January 1980.

Final Report: Advisory Committee on Industrial Innovation. Washington, D.C.: U.S. Commerce Department, September 1979.

Christopher T. Hill and James M. Utterback, eds. *Technological Innovation for a Dynamic Economy.* New York: Pergamon Press, 1979.

National Academy of Engineering, *Industrial Innovation and Public Policy Options: Report of a Colloquium.* Washington, D.C.: National Academy Press, 1980.

3. E. Mansfield, J. Rapoport, A. Romeo, E. Villani, S. Wagner, and F. Husic, *The Production and Application of New Industrial Technology.* (New York: W. W. Norton, 1977). See also the Seminar on Research, Productivity and the National Economy before the Committee on Science and Technology, U.S. House of Representatives, 96th Congress, 2nd Session, June 18, 1980. (Printed report available from the U.S. Government Printing Office.)

4. "Innovation—Has America Lost its Edge," *Newsweek,* June 4, 1979.

5. Ibid.; National Academy of Engineering, *Industrial Innovation.*

6. See Ralph Landau, Chapter 4 in this book.

7. Burton G. Malkiel, quoted in Leonard Silk's column in the *New York Times,* p. D2, October 23, 1981.

8. Harvey Brooks, "Technology, Evolution and Purpose," *Daedalus,* Winter 1980.

9. Ibid., p. 66.

10. William H. Whyte, "Whyte aims 1981 program at reorientation of research," *Footnotes,* Washington, D.C.: American Sociological Association, Vol. 8, No. 6, August 1980, p. 1.

chapter one

SOCIAL AND TECHNOLOGICAL INNOVATION
Harvey Brooks

The following is a slightly edited account of a conversation between a Japanese industrialist and a U.S. Senator concerned with the innovative lag in the United States and the impressive economic performance of Japan. It exemplifies many of the points I wish to discuss in this paper.

Senator: You are right to say there is no short cut to depress the inflation rate of the United States other than to increase the productivity of its industries. But it is not so easy because it requires a vast amount of new investment.

Japanese: Of course there is a way which requires much money. But we have another way which requires only a little money.

Senator: What is the other way?

Japanese: While it may take much money to install automatic machines or robots to increase productivity, improvements can also be made in the way machines are operated which requires only a little money. For example, when you drive a car you may have the frustrating experience of having to stop block after block because of a series of red traffic lights. This stopping would be unnecessary if the signal timing were adjusted to make the lights turn green in pace with the traffic flow. In fact, if this adjustment were made, it might take you only half as much time to reach your destination as before. Thus, merely by synchronizing the signals to the speed of traffic, productivity would be increased at very little cost.

Senator: Oh yes. You mean you can do it by software.

Japanese: Yes. Quality control is a methodology for planning and realizing the most efficient combination of workers and machines . . . I hope you will understand that the control techniques can remarkably increase productivity.[1]

This conversation can be interpreted at two levels. The first is at its face value, namely, that the organization of work and the use of human resources in the work setting is as important for productivity as the technology which is designed to extend the capabilities of the worker. The second may be taken as a paradigm for the whole functioning of a society. It applies not only to the narrow economic objectives of a society, but to its broader social objectives as well. Traffic flow becomes a metaphor for the progress of societal decision making; the traffic lights become the metaphor for the laws and regulations which govern the functioning of the economy. They are much more effective if they are adjusted to the pace of the traffic, if they work with the processes of the economy rather than against them: if controlling the operations of industry does not require continual acceleration and braking with wasteful expenditure of physical or social energy.

Anthony Smith has observed that "innovation consists of skillful development plus social insight."[2] For the purposes of this discussion the essence of the distinction between a social invention and a social innovation is that invention represents the conception of a new set of social relationships based on social insight, whereas innovation is the realization of this conception in practice in the context of real personalities and social resistances. In the past such social insight has often been largely intuitive, but nonetheless important. The great technological innovators of the early twentieth century, such as Henry Ford and Thomas Edison, explicitly considered themselves as social inventors and innovators, even as social reformers. When James B. Conant, chemist and leader of the scientific war effort in World War II, wrote his autobiography, he chose, significantly, to title it *My Several Lives, Memoirs of a Social Inventor.*[3] For an individual whose greatest claim to fame was as a leader of a series of technological revolutions, it is surprising but no accident that he chose to identify as his proudest accomplishments not technical but social inventions. Perhaps nothing better illustrates the fact that great periods of technical innovation in our history, such as World War II, were generally also periods of social invention, and that the entrepreneurs of technical innovation were frequently social inventors and innovators as well. Technical and social invention and innovation go hand in hand, especially when one is talking about what Professor Hughes has called "radical" as compared with "conservative" innovations.[4] Indeed the most important innovation of World War

II was not the technologies themselves, but the unique new relation between scientists and the military which brought the scientists fully into the planning of the tactics for the employment of the new weapons, not just the development of the weapons themselves.[5] This experience also established a new pattern which brought scientists and engineers much more into management and corporate strategy planning in the first two decades after World War II than had been the case in the past.[6]

Similarly, the most important innovation of the Apollo program of the 1960s was probably not the technology which took men to the moon, but the managerial system required to orchestrate the many individual technologies and operational specialties needed for a successful and reliable flight mission.[7] It is difficult to trace the degree to which the lessons from this managerial system have diffused into other parts of our technological economy, since this diffusion has generally not occurred in readily documentable form, but has proceeded by the movement of individuals with experience in the Apollo system. It is also evident that many of the lessons learned in regard to, say, quality control, reliability analysis, and the optimal design of control and display systems for operators have been slow in being transferred to other technological systems where they surely would have been beneficial, most notably in the case of the controls of nuclear reactors.[8] This is a subject to which we will return later.

During the late 1960s the Europeans became very alarmed over what they perceived as the large and growing technological lead of the Americans, which they attributed primarily to massive U.S. government support of high technology development for defense and space activities. The "technology gap" was popularized by the French journalist and politician, Servan-Schreiber, and stimulated a great deal of high-level political attention in the OECD countries.[9] It is rather remarkable that within the short space of ten years this whole discussion appears to have flipped by one hundred and eighty degrees. It is now the Germans and the Japanese that are ten feet tall, and the lag of American innovation is the focus of concern, at least in the United States, though it is not so perceived in Europe.[10] During the years of discussion of the "technology gap" between the United States and the rest of the industrialized world, a few voices spoke out to argue that it was not a technology gap but a management gap. In the words of Theodore Levitt, an American writer on this subject, it was

the great entrepreneurial energy, managerial effort, and involved preoccupation with the consumer's motivations and needs that distinguishes American from European business enterprise. Science is what makes news, especially among men who prefer brains to brawn. But science and advanced technology are not what primarily make the American economy run so fast and well.[11]

This was in mid-1968. Today the same things are being said about the Japanese and German economies vis-a-vis the American, with the tables completely turned. The current American failure is seen as managerial, and in understanding the changing characteristics of both world and domestic markets, and responding in a sufficiently timely manner to these changes.[12]

All of these examples illustrate in different ways the importance of the social dimensions of invention and innovation. While it is certainly an oversimplification to ascribe comparative economic performance either to technological or management factors to the exclusion of each other, it is clear that both dimensions are important. Solutions that focus on only one dimension are not likely to be productive. What is uncertain, of course, is the principal direction of causation—whether it runs from managerial and social failures to technological innovation, or whether it runs from technological lag to managerial conservatism. Undoubtedly, there is a high degree of synergism between the two, and this is what makes many debates over the causes of the current American syndrome somewhat pointless.

The Innovative Malaise

There is no doubt as to the widespread perception of a lag in innovative capacity in the United States. The consensus that there is a lag in performance in the U.S. economy as compared with our main competitors, Germany and Japan, is almost complete. It applies both to competitive performance in international trade, and to indices of performance of the domestic economy, such as the rate of growth of GNP per employed person, output per manhour, or total factor productivity.[13] Beyond this, however, there is almost no consensus as to the causes of the gap or its true significance. In the first place, the gap is in the rate of change of the indices, not the absolute indices, which still place the United States ahead of its competitors in most sectors. Thus the question can be raised as to whether we are dealing with a catch-up from a lag originating in the destruction of World War II, or whether we are on the verge of being overtaken by our rivals. Some writers have suggested, for example, that the new generation of Japanese who have been brought up in comparative affluence do not have the same devotion to work and productivity as their parents. Furthermore, it is pointed out that the Japanese ''economic miracle'' is restricted to a few modern industrial sectors, and that much of the small business, distribution, and service sector—which is less exposed to foreign competition—is inefficient, disorganized, and overmanned, and also very vulnerable to even a small slowdown in economic growth. There is a question as to whether both the Japanese and German miracles do not harbor the seeds of their own decay.[14]

In addition there are increasing questions about the true significance of productivity statistics, especially in the light of rapidly changing structures of the economies being compared.[15] All the OECD economies have experienced a major reduction in productivity growth after the 1973 oil crisis compared with before the crisis. This has led to the suggestion that a prime cause of the decline is the increased cost of energy,[16] which has led to the substitution of labor for both capital and energy. Another favored source of stagnation is the explosion of health, safety, and environmental regulations. Nevertheless, careful analysis indicates that this can at most account for 8 to 12 percent of the decline in productivity growth in the 1970s, and there is some indication that this percentage is now on the decline as we overcome the lag in standards that most observers agree existed prior to the late 1960s.[17] Other culprits include the changing average skill level of the labor force (higher entry rates of youth, and inexperienced women, and minorities), reduced rates of savings and capital investment, and reduced expenditures for economically oriented R & D as a fraction of GNP.[18]

Perhaps the most important generalization that can be made from all this is the danger of overgeneralization. Technological innovation is a highly individualized and idiosyncratic activity, particularly when we focus on the most radical and economically significant innovations. The performance of the U.S. economy is highly differentiated among different industries, and the lag in innovation does not apply across the board. Public concern has been prompted by highly visible crises in selected industries, notably automobiles, steel, and color television sets (as well as other consumer electronics). Several of the most ailing industries—steel, shipbuilding, the merchant marine—suffer from worldwide overcapacity, and are also in some trouble in the countries which are our major competitors.[19] The U.S. performance in its traditionally strong sectors—aircraft, advanced weapons systems, computers, telecommunications, industrial electronics, heavy machinery, and agricultural goods—remains strong and growing.[20] Automobiles represent a very special situation brought about by a variety of factors which include failure to commercialize known innovations important to future markets, but misguided government energy policies must also bear a large share of the blame.[21] There is room for doubt as to the degree to which wiser corporate strategies could have avoided the present plight of the

American auto industry, given the wild gyrations in the composition of domestic market demand which have occurred since the late 60s (at least in part induced by misleading price signals to the American public resulting from government policies).[22] The question remains whether, given the economies of scale characteristic of automobile production and marketing, the situation has already become irretrievable as a result of the Japanese headstart in the world market, at least without a major injection of new capital and restructuring of the industry.[23]

With respect to the areas of U.S. continuing leadership, on the other hand, there is also significant debate as to the permanence of the lead which all acknowledge to exist at the present time.[24] In modern information technologies the U.S. lead is being severely challenged by the Japanese with considerable government assistance.[25] The Europeans are attempting to challenge the Americans in civil aircraft with the Airbus, although this uses more than 40 percent American components, including General Electric turbofan engines. There is some doubt whether either the Europeans or the Japanese are close to matching U.S. productivity levels in aircraft production.[26] The Europeans and Japanese are beginning to challenge the United States by developing independent satellite launch facilities, but this would have important repercussions for the U.S. lead in space technology generally only if the Space Shuttle should prove a failure.

The French are offering services in a new remote sensing satellite SPOT projected for launch in 1984, while the United States is still debating whether and how to place its remote sensing capabilities on a commercial footing.[27] The French are already well ahead of the United States in breeder reactor technology,[28] but there is a major debate as to whether they will be successful in commercializing the technology or can capture a significant part of the world market should the technology prove to be economically competitive with other energy sources. Electricity from breeders may not be competitive with that from U.S. designed light water reactors until the price of uranium nearly doubles over its present value. With reduced projected demands for energy this may be several decades away. The United States may actually be better off with a follower strategy, much as it was with jet aircraft in following the British Comet in the 1950s. The United States appears to have a lead in the new biotechnologies at present, although there is now a worldwide race to exploit rDNA techniques and the Japanese have the advantage of a very well developed fermentation industry which might take advantage of genetic engineering for industrial purposes.

However, it is too early to draw conclusions regarding the most advanced technologies. During the 1950s and 1960s the Europeans, especially the British, could be said to have kept abreast of the United

States *technically* in most of the technologies in which we later achieved a commanding *commercial* lead. The proof of the pudding will be in commercial, not technological, success, and here questions of timing, management, and marketing will be crucial. The "managerial gap" which seems evident in the older industries such as steel and automobiles does not appear to exist in the technologically most advanced industries.[29]

On the other hand, the newer technologies may not enjoy the same degree of synergism with space-defense requirements that was the case with aircraft, electronics, and nuclear reactors, and which may have been an important source of American commercial advantage. In summary, there is presently considerable disagreement as to whether the American advantage may be eroding across the board, so that its advantage in the high-technology sectors is a transient one, or whether our advantage in these sectors will persist. The answer will probably depend to a large degree on how well social and managerial innovation continue to accompany technological innovation in the American context. The "product cycle" theory of comparative advantage suggests that the United States must continue to pioneer the most advanced technologies. Being inherently a high labor cost country because of its high standard of living, the United States will not be able to retain its comparative advantage in a new industry as the technology matures, at least if it is to retain its advantage in living standards overall.[30] Thus it must continually innovate with new products and services, often replacing maturing industries, if it is to maintain simultaneously a superior average standard of living, a favorable trade balance, and full participation in an open international trading system. To some extent the decline in the maturer industries may be arrested by superior managerial and marketing strategies and hence by continuing social invention and innovation.

The present economic decline of the United States depends to a high degree on the rapid change which has taken place in the general economic, social, and international environment in the last decade. The change has been considerably more rapid for the United States than for its principal competitors, who had been more accustomed to operating in an environment with constraints on the supply of natural resources and energy. Consequently our competitors were better positioned to move ahead in the new circumstances. For a century the U.S. economy had grown in circumstances of abundant resources and energy and relatively scarce and expensive labor. By and large the situation in Europe and Japan was just the opposite.

Moreover, the United States had followed a deliberate policy of subsidizing both the production and consumption of energy: production through such devices as the oil depletion allowance (which also applied to other nonrenewable resources), and consumption through such devices as public financing of highways, airports, and other transportation infrastructures as well as tax advantages for single family housing which favored suburban spread. By contrast, the Europeans and Japanese discouraged the expansion of personal transportation through high taxes on gasoline and taxes on automobiles proportioned to their fuel consumption. Government policies also favored public transportation, rail over air, and compact cities. Even industrial energy was more expensive abroad so that generally industry had an incentive to be more energy-efficient than in North America.[31]

Although these differences were declining with time, European and Japanese industry was still better positioned to meet the new market situation of the 1970s, particularly in the case of automobiles. The situation in the United States was exacerbated by three other trends which occurred simultaneously. These were:

1. The very rapid rise in expectations for a clean environment and higher standards of health and safety in the workplace. Although this trend occurred almost simultaneously in Europe and Japan, it lagged behind the United States in respect to rate of change, both because expectations grew more slowly and because standards had previously been generally higher. In fact, America was playing "catch-up" in both environmental and social legislation in the early seventies.

2. The new international division of labor brought on by the rapid economic growth of Europe and Japan as well as a few rapidly industrializing LDCs such as Brazil, South Korea, and Taiwan. This situation was exacerbated by the fact that many foreign manufacturing facilities were of more recent vintage than the American ones, and therefore more efficient in terms of both energy and, in some instances, labor use.

3. The revolution of entitlements. Again this was a worldwide trend, but one which proceeded more rapidly in the United States in part because of a lag prior to 1970 and in part because expectations were higher.

All of these environmental changes in the economic environment hit an economy overconfident of its own capacities, a management confident in its superiority, and engineers used to being at the head of the technological procession and who felt they had little to learn from others, along with something of a "not invented here" complex. As a result American business and management generally were much less inclined to scan the world pool of management techniques and technology for ideas that might be used to advantage. By contrast, the Japanese had a long tradition of playing a follower strategy, and were unusually adept at scanning the world for new ideas and adapting and using them in their own circumstances, with no nonsense about pride of authorship or originality. The long habit of success has died slowly in the United States, and the long habit of learning from others has stood our competitors in good stead. Exposure to world competition is a relatively new experience for U.S. business, especially for those industries which are now in deepest trouble. The natural tendency of the more mature industries to lose their comparative advantage has thus been exacerbated by the managerial mindsets induced by the habit of success during the long postwar period when they had little effective competition from a world still recovering from the ravages of war.

The Social Dimensions of Invention and Innovation

We now return to the main theme of this paper, the importance of social invention and innovation in the overall process of technological innovation. Hannay and McGinn have stated that the "basic function of technology is the expansion of the realm of practical human possibility."[32] In itself this definition implies a good deal more than physical artifacts, and hence a social dimension of technological innovation. The wide adoption of a new technology almost always involves the simultaneous creation of managerial and social supporting systems to sustain its use.[33] The proper conception of such supporting systems may be more important to the commercial or social success of the innovation than the physical invention itself.

We will now consider a categorization of types of inventions and innovations according to the importance of their social dimensions. The types are pure social inventions and innovations, sociotechnical system inventions and innovations, and pure technical innovations. Actually, we shall see that in practice there are no entirely pure types, but rather they blend into each other in an almost continuous distribution.

Social Inventions and Innovations.

Some examples are: withholding taxes, no fault insurance, the supermarket, McDonald's, the United Parcel Service, environmental impact statements, the Cooperative Extension Service of the Department of Agriculture, health maintenance organizations, public utility holding companies, government research contracts, the negative income tax, enterprise zones. None of these is purely social or managerial; they almost always draw in some ancillary physical technology. Withholding taxes would not have been practical on the present scale without the

modern computer. The supermarket has resulted in the invention of
new types of check-out counters, stackable grocery carts, optical labeling
of cans for automatic check-out, etc. McDonald's developed a whole
host of minor but important inventions such as a special scoop and bag
for french fries.[34] The thrust, however, comes from the market, and the
technology is usually incidental and rather mundane in technical terms,
though no less ingenious. The organizational invention comes first, and
technical innovations are gradually introduced to improve it, rather than
the reverse.

Sociotechnical Inventions and Innovations.

These are essentially clusters of innovations, often accumulated around
a single central artifact such as the automobile, the airplane, or the
television receiver. Yet these artifacts have little or no social value without a large system of ancillary technologies and social inventions and
innovations. The automobile is part of a system which includes highways, filling stations, repair shops, dealer agencies, traffic controls, the
insurance system, the legal and judicial system associated with automobile accidents and traffic violations, the licensing system, the driver
training system, advertising, and so on almost endlessly. This is without
mentioning the whole supplier system of subcontractors and parts manufacturers. Of course the supporting systems themselves require much
technological innovation from metered gasoline pumps through automatic or programmed traffic signals to radar speed traps and citizen
band radios.

Similarly the airplane system includes air traffic control, radar location
of aircraft, navigation systems, airline reservation systems, a weather
forecasting and observation system, airports, baggage handling systems, loading ramps, certification system for airworthiness, meal services, an elaborate and highly reliable aircraft maintenance system, etc.

The case of TV is of special interest because, in contrast to aviation
which has become standardized worldwide, it has evolved different
supporting systems in different countries. In the United States the system grew up with advertising support, more recently supplemented by
public television and various forms of pay television. In Britain the system became a public monopoly supported by license fees on individual
receivers, more recently supplemented by an advertiser-supported commercial system. Still other schemes evolved in Germany and France.
New supporting systems and managerial arrangements are now evolving with the introduction of cable TV and soon-to-be-introduced direct
broadcast via satellite. All of this took or will take a great deal of social

innovation scarcely imagined at the time of the original invention. Nor is the present system the only one that could be imagined which is compatible with basically the same physical technology. Much of the supporting system depends on the system of frequency allocation which was established very early and is international. One can imagine other alternatives,[35] just as one can imagine a system of personal transportation based on rental rather than ownership, as with the telecommunications systems, early computer systems, and xerography.

The variety of possible sociotechnical systems is greatly enhanced with the introduction of many new information technologies, which for the first time offer the opportunity to convert many highly labor-intensive service delivery systems into more capital-intensive sociotechnical systems. Examples are offices, financial systems, insurance, tax collection, welfare, medical records, and even education. An example of a mature sociotechnical system of this sort is the telecommunications network, now rapidly ramifying into a data management and processing network of international scope. A sociotechnical system at a much more primitive stage of development is the computerized office. Still another sociotechnical system which is fairly mature is the whole system of newspaper publishing from reporting and news gathering to final delivery of the individualized newspaper. This system has evolved dramatically over the last decade, especially in the United States, but generally throughout the world. It is often said that up until the end of the 1960s the technology of printing and publishing had changed very slowly since Gutenberg, but since then has been completely transformed.[36] Yet this transformation is probably still only in its early stages.

With the costs of transportation and newsprint constantly rising because of energy costs while the costs of communications and data management are constantly declining, it is probably only a matter of time before the economic advantage in publishing shifts from print to electronics. One observer has estimated that "if the price of gasoline rises to $1.50 per gallon, the whole operating profit of a newspaper would disappear."[37] The transformation of publishing has involved as much social invention and innovation as technical innovation. In the introduction of computerized methods into the newsroom, "many of the principal pioneering newspapers have had to undertake detailed descriptions of themselves as communication and social systems while preparing the specifications for the physical machinery they are ordering."[38] Moreover it has been observed that the most effective experimenters "were those who employed democratic or highly consultative processes of planning."[39] The same considerations will undoubtedly apply to the coming revolution in office technology. The social processes followed in the introduction of new technology, as well as thorough understanding of the human systems and interrelations involved, will be especially crucial to successful innovation in information technology.

Innovations Nearly Purely Technical

There are many areas of innovation where change appears to be more purely technical, though never exclusively so. This is probably particularly true in the case of intermediate inputs, goods traded between firms as inputs to the production process itself. A prime example is the area of materials. Materials with new properties have been a major source of innovation in both products and manufacturing processes in the last two decades, and this is likely to continue into the future as changing factor and raw material costs and politically insecure sources of supply stimulate the substitution of materials as well as the recycling of materials.[40] Most materials innovations are introduced with little consideration of sociotechnical aspects, although they sometimes have a big effect on the organization of manufacturing processes, e.g., when plastics are substituted for metals and alloys.[41] Radical innovations such as the transistor, the laser, or integrated circuits usually begin as purely technical inventions, but as they draw in ancillary technologies and are incorporated in manufacturing processes the more social aspects frequently become apparent. This is especially true when quality control and production yields are crucial to cost.[42]

The question of pure technological innovation versus sociotechnical invention and innovation also relates to the argument over the relative importance of market pull and technology push in the generation of innovations.[43] Generally speaking, pure technology innovations are those that arise directly from scientific discoveries, and are frequently of the "radical" type described by Hughes. Nevertheless, the exploitation of such pure technology inventions usually ends by requiring both ancillary technology and sociotechnical adaptations before the invention is used on a truly significant scale. For example, although the laser can be regarded as an almost "pure" technological invention, its incorporation into an optical communications system has entailed a whole cluster of important ancillary technologies including low attenuation optical fibers and various kinds of optical processing devices. The laser is an ingredient in literally thousands of inventions: precision surveying equipment, earth strain sensors used in connection with earthquake prediction, precision alignment devices for road-building or airplane manufacture, processing devices for metal surfaces, medical devices for repairing detached retinas, tools in high precision microchemical analysis (laser plume spectroscopy), devices for remote monitoring of atmospheric pollutants, the basis of a new method for isotope separation, and so on almost indefinitely. Most of these applications can be considered to have been generated by market pull once the properties of the various basic types of lasers had been explored and described.

When innovations are described as "purely technical" we mean only that the conception of the device or process derives entirely from a scientific discovery or from earlier technology rather than from the perception of a social need of a potential market for an application. We do not mean to imply that what is discovered and when it is discovered is not influenced by the cultural milieu in which the innovation arises or by the organizational setting in which research and development are carried out. It is obvious that some types of cultures and certain types of social character or personal background are more likely to give rise to original discoveries or inventions than others. But this is a much more subtle effect than a response to signals from society as to what it needs.

The three categories of innovations we have tried to classify above are all of equal importance and frequently interact with each other. As we have seen, pure technical innovations create a situation which generates new social pulls for other innovations which incorporate the original technology, eventually into a new sociotechnical system. Similarly, novel organizational arrangements frequently elicit technological innovations.

A Typology of Social Invention and Innovation

In what follows we will classify social inventions and innovations into four types:

1. Market inventions and innovations
2. Managerial inventions and innovations
3. Political inventions and innovations
4. Institutional inventions and innovations.

Market inventions and innovations are new ways of marketing services or products which permit a greater expansion of the market for existing technologies or make possible the market penetration of a new technology. They may interact with political innovations in that a market is facilitated by a political reform. For example, the enormous expansion of owner-occupied housing after World War II in the United States was the result of a series of political reforms that restructured the capital market for housing and made possible an enormous increase in housing demand. This has been described as a "cornucopia of federal benevolence aimed at making America a nation of homeowners."[44] Despite pockets of substandard housing, it has made the average U.S. citizen the best-housed citizen in the world by quite a large margin.[45] But market innovations may also consist of purely private devices such as quick food service, the renting of computer and copying services, automobile rentals, or TV advertising.

Managerial inventions and innovations will be defined as changes in work organization within firms or bureaucracies which improve productivity, the quality of products or service, or simply the quality of working life. As we have already suggested this is the area which is probably most responsible for recent Japanese competitive success in world markets.

Political inventions and innovations are new government policies or legislation designed to attain new social objectives. They usually also involve the creation of new governmental institutions; however, the effectiveness of political innovations is most usually measured by some change in the demand or supply of some socially valued good or service, what Musgrave has labeled "merit goods."[46] Two examples might be Medicare and Medicaid, which changed both the demand for and the distribution of health care services and hence the health care market in the United States. Another example is the explosion of environmental legislation in the 1970s which has created a market for new technology in environmental monitoring and control which is still growing rapidly.

By *institutional inventions and innovations* we refer to new kinds of institutions designed to provide new services or social functions. Examples are health maintenance organizations (HMOs) which deliver prepaid medicare care,[47] and various kinds of quasi-private or public-private corporations designed to deliver various social services.[48] Institutional inventions and innovations may overlap with the three other kinds. For example, HMOs could be regarded as a new form of marketing for health care, and their development has been considerably stimulated by federal legislation. Similarly many novel quasi-private corporations in the social service field were developed as a response to governmental "great society" legislation.

Market Invention and Innovation

Many writers on technological innovation have stressed the importance of social insights in envisioning what society will need and want in the future. Except for relatively minor improvements in existing products and services, people seldom know what they want until they have a fairly concrete idea of what is possible technically and what it will cost. Tremendous importance thus attaches to demonstration effects, which are most frequently provided first in affluent or institutional markets.

In the important fields of telecommunications and data processing the market has evolved through a hierarchy of uses beginning with military space technology, then extending to business demand, and finally to the individual consumer. It seems most likely that if various solar energy technologies ever experience significant market penetration it will be the result of the demonstration effect of purchase by affluent consumers as prestige items. Note, for example, that an early major commercial application of solar collectors has been for the heating of private swimming pools.

A generic class of marketing inventions and innovations likely to be important in the future is the substitution of the marketing of services for the marketing of products. Probably a major reason for the great success of the telephone industry, the computer industry, and the copying industry was the fact that what they initially marketed was services rather than equipment. Thus the customer did not have to gamble on the purchase of an expensive piece of equipment whose value to him was difficult to test in advance. He paid for services only as used while the supplier carried the risks of ownership. In the future we may see the development of markets for transportation services or energy services where the seller similarly owns the equipment and charges the customer only for the service provided when used.

This is particularly well illustrated in the case of energy, an intermediate good which is valuable only for the services it provides—comfort in the home, personal transportation, mechanical energy, or process heat for the transformation of materials and the fabrication of goods. As energy has become more expensive, there is increasing realization that the same services to consumers can be provided with much less use of primary energy resources than is needed in present systems. With current technology homes can be built that use only one-fifth the amount of fuel of the average dwelling in the 1970 housing stock; in a few years the automobile fleet will be more than twice as energy-efficient as the 1975 fleet; the energy intensity of manufacturing has already declined by more than 10 percent. Roger Sant[49] has projected that over the next twenty years, by following a least cost strategy, the cost of energy services *per capita* could be less than it is today.

In other words the improvement in end use efficiency which could be achieved by substituting more expensive but more efficient equipment for present energy-using equipment would slightly more than pay for itself. By the year 2000 the total value of primary energy consumed *per capita* would be down about 30 percent with the aggregate capital cost up by about the same amount, assuming recent Energy Information Agency projections of future fuel prices. This projection is most likely to be realized if the energy industry is gradually converted to an energy service industry which competes to provide defined energy services at minimum cost. Thus, for example, a utility might provide capital for an energy efficiency investment instead of simply selling energy.[50] The resulting technological system of energy delivery might become much more diversified in character than present networks, though this is by no means certain.

Similarly, in personal transportation, some observers have envisioned a very rapid growth of rental markets, with a much more diversified mix of special purpose vehicles designed for various trip lengths, average speeds, and passenger capacities.[51]

With respect to computer-based services the real revolution may be yet to come and will be paced more by ingenious marketing inventions and innovations than by technology. Throughout the industrialized world there is already extensive experimentation with various home computer-based services such as Prestel in the United Kingdom,[52] the Antiope system in France,[53] the Captain system in Japan,[54] the German Bildschirmtext system,[55] and many other systems such as Didon, Oracle, Qube, etc.[56] Each of these systems is different as a result of the different complex of laws, customs, rate structures, and vested interests that exist in each of the countries involved. None has yet reached the stage of a self-evidently viable mass market, and it will probably be at least a decade before the viability and probable structure of such a market becomes clear. As much creativity and imagination are likely to be required on the marketing side as in the hardware, although the two will doubtless interact in an iterative manner.

Managerial Invention and Innovation

Most observers now agree that the remarkable competitive performance of the Japanese economy is due more to managerial than to technological invention and innovation, even though the Japanese have also shown great skill in adapting their technology to meet the special requirements of foreign markets. Interestingly enough, managerial success in Japan has been largely on the production side. They have probably been more successful in penetrating foreign markets than they have been in selling in their own domestic market.[57] One could argue, in fact, that in the case of automobiles Japanese success may be explained by the marriage of a very efficient Japanese production system with a very efficient American marketing network already in place for domestic U.S. automobiles.

Perhaps the most important Japanese social invention is a labor system which makes Japanese workers much more willing and even eager to accept new production technology and to participate actively in the process of ascending the learning curve of manufacturing productivity.[58] There is a good deal of disagreement as to how much of the Japanese management philosophy is actually transferable to the American scene. It is sometimes asserted that the Japanese system of labor management relations, involving lifetime employment, an age-graded wage structure, and enterprise unions, is peculiar to Japanese culture and historical development and hence not readily transferable to the American context. On the other hand, Japanese analysts have argued that their system is a rational economic adaptation to the particular structure of the Japanese economy and labor force age structure during the 1960s and 1970s—rapid economic growth and high investment rate, a young and inexperienced labor force, and a large reservoir of temporary workers.[59] According to this theory, profound changes are in store for Japanese labor practices as the economy faces slower growth, an aging work force, and a static or declining population with high levels of education. The present Japanese system may be of only limited applicability to the United States, where the structure of the economy and the labor force have been more similar to what Japan will move into in the 1980s.

Nevertheless, there is now sufficient experience with management innovations in the United States to suggest that rather spectacular increases in productivity can be achieved with rather modest inventions and innovations in the organization of work in both manufacturing and service industries. This can occur with a very low investment either of capital or of the time of management and production workers. For half a century the American manufacturing system has been dominated by the philosophy of Frederick Taylor[60] and this philosophy of "scientific management" appears to have served us fairly well in the social and cultural context which existed up until about 1950 or so. This tradition could be best described as one of technology being "the irresistible prime mover that inevitably defines working tasks and working environments."[61] Increasingly, however, it is becoming necessary to replace this philosophy by one of "sociotechnical design," the matching of the human and technological system, not by forcing humans to adapt to the imperatives of a pre-designed technology, but by mutual accommodation of human and technical aspects which fully respects the aspirations, personal dignity, and self-development of all the people engaged in the process from the lowest to the highest.[62] It almost certainly involves a departure from rigid hierarchies and greater involvement of all levels of the work force in the planning of the work process. The paradox is that if one attempts sociotechnical design with improved productivity as the sole objective, the process usually fails. The productivity improvements that have appeared with managerial innovation seem to have come out primarily as a by-product of efforts to improve the human aspects of work organization by other than pure efficiency criteria.[63] Yet improved productivity is not an automatic consequence either.

There are several reasons why management inventiveness and innovation is necessary in the contemporary social context.

1. The modern work force has higher expectations conditioned largely by higher educational levels and increasing democratization of other aspects of modern society.

2. The nature of jobs, especially in the newer types of industries and the rapidly growing service sector, is such that productivity is much more dependent on voluntary cooperation and high work commitment on the part of all employees. In more and more work situations the costs of policing employee performance are prohibitive, and productivity must depend on a high level of mutual trust between management and workers. Such trust can exist only if there is an expectation on both sides, based on concrete experience, that cooperative behavior will not be to the disadvantage of either and will in fact be of mutual benefit. On the one hand, the economic benefits of any productivity gains must be shared between workers and management in a manner which both sides can see as equitable. In particular productivity gains cannot threaten job security. On the other hand, management support for human development cannot be taken advantage of by workers, since this not only undermines management trust in workers but also undermines the sense of fairness as between workers who cooperate and those who "goof off."

3. Managerial invention and innovation under present conditions probably has higher returns, relative to the effort expended, than pure technological innovation in many cases. This is in part due to the increasing cost of capital, but also due to the fact that the flexibility of modern production and office technologies offers much greater opportunities for varying organizational designs to improve the human environment and realize the full human potential of the organization. Indeed the recent past lag in organizational invention and innovation has created many unexploited opportunities for matching human and machine potential while improving the quality of work.

Political Invention and Innovation

As indicated earlier, there is frequently a close interaction between political invention and innovation and markets because a major purpose of political innovation is to compensate for the failure, or alleged failure, of markets to result in desired social outcomes. A type of political innovation that is attracting increasing discussion is designed to make markets work better, rather than to compensate for or offset market outcomes. This type of innovation is designed to change the basic rules of the market in such a way as to make the profitability or growth of firms more nearly a reward for activities which further the public interest by making private benefit more congruent with public benefit. This usually entails the replacement of standards and prohibitions by economic incentives—for example effluent fees or marketable emission permits.[64] Two recent examples of political inventions are the introduction of the "bubble concept" for regulating emissions by EPA[65] and the recent "superfund" legislation for dealing with past hazardous waste dumps.[66]

In theory, at least, the bubble concept operates somewhat like a marketable pollution permit in that a firm can place a new polluting activity in an area where the maximum concentration of a given pollutant approaches allowable levels if it can subsidize another firm already in the area to reduce its emissions sufficiently to more than offset the additional pollution caused by the new firm. In the case of the superfund legislation the original producer of a chemical which is ultimately a source of potential hazard somewhere down the line is made responsible for any hazard which is traceable to its product even if it has been transformed many times by subsequent purchasers of the material. This provides a strong economic incentive for the original producer to worry about all the subsequent uses of his product and its ultimate disposal. We are not arguing here whether this is a desirable or fair proposal; from a public policy viewpoint it may or may not also create too much of a disincentive to chemical innovation. But it *is* a new political concept of wide generality and can be regarded as a political invention which strongly affects market incentives.

In no area is the need for political inventiveness more apparent than in the area of energy policy. In part this is because energy policy is already so politicized in and among the industrialized countries, and between the industrialized world and the rest of the world. Thus even solutions to the energy dilemma that rely heavily on market forces will require imaginative political invention and innovation. In the United States, as elsewhere, a large part of the energy industry is regulated; this is especially true of the electric utility and gas utility industries. A regulatory system that worked relatively smoothly in a period of declining prices and increasing economies of scale in production and distribution has now broken down. One example of this is that a regulated rate of return based on historic capital costs for both electric and gas utilities leads to prices to consumers that are less than the incremental costs of additions to supplies. This is a prescription for overconsumption and underproduction and can only lead ultimately to a production capacity which falls short of demand at the existing price.[67]

Because of the long lead times involved in the installation of new production facilities, however, the true situation tends to be obscured. Since 1973 it has been further obscured by the use of the fuel-adjustment clause which allows utilities to pass through the full extra cost of more expensive fuel to consumers and thus removes any economic incentive to replace uneconomic oil-fired generating facilities with alternative coal, nuclear, or other types of plants, with lower fuel but higher capital costs. In addition, regulation often discourages the integration of decentralized generating facilities, such as small-scale hydro and cogeneration into the electric grid even when it would be economically advantageous to do so. The new PURPA legislation requires utilities to buy back power from such decentralized sources at a price equal to the price of the power that the utility would otherwise have to generate itself. This is a simple concept in theory, but requires a tremendous amount of imaginative legal and regulatory inventiveness to implement in practice.[68] Similarly regulatory innovations and inventions are required to make it attractive and possible for utilities to provide capital for energy efficiency investments where the net cost of the energy saved (taking into account future price increases) would be less than the cost of additional energy production. There is a good deal of experimentation in the states with such innovations now; what is needed is careful monitoring of the operation of some of these laws to assess their effectiveness in influencing the flow of capital into energy efficiency investments.

What may be more important than specific political inventions and innovations that affect markets in regulated industries is the development of a more systematic process for them in this area. In fact there is very little institutionalized invention and innovation in the political field analogous to the development process for hardware. In particular there is no systematic feedback of experience from actual attempted political inventions and innovations to evaluate their effects. Moreover, it is very difficult to carry out localized experiments. To some extent the individual states can act as laboratories for energy policy development. For example, Oregon, Vermont, and several other states have experimented with modified rate structures including time of day and seasonal pricing for electricity in order to reduce peak demand. Several states, including Oregon and California, have also developed legislation to permit or compel utilities to finance energy conservation investments or renewable energy sources and to provide tax benefits to consumers for such investments. Much could be learned from systematic studies of the effects of such initiatives on energy demand, load profiles, and fuel mix as well as economic effects such as income distribution. It has been proposed, in fact, that the federal government find ways to support carefully designed localized pilot experiments in the field of energy conservation and environmental protection analogous to what was attempted for the negative income tax, housing, and school vouchers.[69] This could be regarded as social research, development, and demonstration.

Perhaps the greatest need for political invention and innovation lies in breaking out of the adversarial relationship that is now paralyzing decision making in the United States in most areas related to the development and deployment of technology, particularly as they impact on health, safety, and ecological effects. It is not only the long dragging out of decisions but also the tremendous consumption of intellectual talent and political skill involved in fighting through these issues.[70] Our political and legal system appears to be losing its capacity to integrate conflicting interests into a sustainable program of action for dealing with many of our major problems.[71] At the moment other industrialized societies, notably Japan and France, seem to be more successful in this, but even in these cases the greater apparent coherence of national policy may be as much a function of the relative political impotence of certain interests and political perspectives as of superior political technology for building national consensus.

The pluralist and decentralized American system seems well designed to provide *de facto* veto power to any interest group that feels sufficiently challenged by a national policy. On the one hand, environmental interests have effectively brought the nuclear power industry to a halt.[72] On the other hand, administrative and judicial due process have been effectively used by some industrial interests to delay or frustrate many environmental regulations.[73]

Various modes of public-private collaboration in establishing agreed-upon goals have been proposed or tried. One example is the Health Effects Institute set up cooperatively between the automobile industry and the EPA.[74] Another is the national Ocean Margin Drilling program (OMD) established as a collaboration between the National Science Foundation (NSF) and the oil industry.[75] The processes used by the Office of Technology Assessment (OTA) of the U.S. Congress in reaching out to a broad diversity of both interest groups and expertise in developing credible objective analyses of important national issues involving science and technology is another example of a promising political innovation.[76] The early organizational troubles of OTA have delayed the possibility of a realistic evaluation of the promise of this mechanism. A number of private sector initiatives for bringing together conflicting interests to arrive at a partial consensus on environmental

policies have been developed with governmental and private foundation support. Two interesting examples of this are the Nuclear Waste
Management Process Review Forum developed by the Resolve Center
for Environmental Conflict Resolution[77] and the National Coal Study.[78]
The difficulty with these mechanisms is that, although they provide a
good forum for the representation of all significant interests, the interest
groups are not bound by the conclusions reached by their representatives in the forum, and frequently embrace quite different positions in
the more adversarial atmosphere of the courts and official administrative
hearings. The best hope is that cumulative experience with such nonofficial dialogues, combined with public frustration over the paralysis
of decision making, will become a learning process which will ultimately
carry over into the more formal political and legal forums.

There is apparent a growing bipartisan public consensus in the United
States that the historical processes of consensus building that have
served us well in the past are not working today and are leading to
confrontation of fragmented interests at higher and higher levels of
government and the courts. There is also an increasing realization that
the country cannot resolve its present dilemmas without at least temporary material sacrifices on the part of most of its citizens, sacrifices
entailed by the necessity to channel more resources into investment in
future productivity and resource development and into replacement of
a decaying and obsolete infrastructure in such areas as transportation,
urban water and sewage, and waste management facilities.[79] In a democratic polity these sacrifices cannot be imposed, at least for long; rather
they must be negotiated in a social process which penetrates to the grass
roots.[80] Such a negotiation can only take place when there is a widespread realization among all the interests involved that sacrifice is necessary, and that government can no longer be used as a lever to avoid
the burden on one group by pushing it onto others. It is this realization
which now appears to be growing, and may provide the necessary political climate for experimenting with inventive political mechanisms for
conducting the negotiating process.

Institutional Invention and Innovation

The relationship between technology and social institutions is a subtle
one. Neither fully determines the other. They evolve partly independ-
ently of each other, and partly through an iterative sequence of interactions. These interactions occur from the microscopic level of the single
office to the macroscopic level of the whole economy or polity, and at
all levels in between. The interesting issue for contemporary discussion
is the extent to which these interactions are predictable and hence con-

trollable. Technology evolves under the influence of social and economic selection, but successive generations of technologies have genetic relationships which are inherent in the technologies themselves rather than the social milieu in which they reside.[81] Thus technological evolution is like biological evolution, driven by a subtle combination of technical "inheritance" and socioeconomic selection. Certain institutional or social settings make certain technological choices more probable than others, but certain configurations of technology make certain institutional choices more likely than others.

These concepts seem particularly relevant to the prospective introduction of new information technologies into offices, for example. It has frequently been pointed out that these technologies make possible the Taylorization of office work, changing the office from a somewhat craft-like job shop to an intellectual assembly line, with loss of control by the individual over the pace and independence of his or her work and consequent loss of self-respect and job satisfaction. Yet this is probably more a function of the preexisting assumptions of management than it is inherent in the technology itself. The technology merely permits the carrying of this mindset to its most extreme logical conclusion. At the same time, since information technologies tend to be modular and their use in practice tends to be determined by how the equipment is programmed, this flexibility of "software" provides unparalleled opportunity to restructure the office according to more "human" criteria of organizational design, allowing for greater individual autonomy and control of working pace and social interactions. However, the seizure of this opportunity may require much more conscious and early attention to the redesign of social interactions in the office than has been customary in the introduction of new technologies into work settings. The challenge of organizational redesign is made more difficult by the fact that office technology is usually designed and programmed by vendors and in-house specialists whose incentives come from the sale and rapid debugging of equipment rather than from the quality of the services ultimately rendered by the office.

Since the quality of human interaction in the office is essential to its effectiveness and productivity in the long run, the people who will actually operate the system should be closely involved in the process of system design, which will have to be at least partly a negotiation between the various participants in the design process and the users.

The flexibility of the information technologies which provide the theoretical opportunity for innovative organizational design is probably characteristic of many new technologies that will be introduced in the next two decades. Thus the opportunities for social invention and innovation at many levels can match the need for it generated by the rapid changes taking place in the world social and economic environment. At the same time new technologies also provide greater opportunities for freezing in obsolete social practices and managerial concepts, as is illustrated in the "Taylorization" of offices.

The category "institutional invention and innovation" that we have used in this section is differentiated from market, managerial, and political invention and innovation primarily by its scale. It is intended to apply at the "micro" level, i.e., at the level of the quality of social interactions at the working level and in individual lives rather than at the level of larger institutions and formal governance mechanisms. Obviously this cannot be a sharp differentiation, for large organizations cannot work well in the long run unless the quality of human relationships at the working level is fulfilling to the individuals involved. This in turn also depends on agreement on common goals among larger groups and on the process by which such agreed-upon goals are negotiated in the overall organizational setting.

The last two decades have witnessed a remarkable growth of political regulation of private and local activity by higher levels of government and by the courts. This intervention ranges from environmental, health, and safety regulation and enforcement of strict liability in the courts to the mandating of special treatment or social support for various groups in society whose interests or views had not previously been taken into account in decisions at many levels from corporate to governmental, even when they might be affected, directly or indirectly. One could describe the evolution of the last twenty years as an attempt to enforce a wider set of criteria on decision makers in all the institutions of society—wider in the sense of taking into account considerations less evidently related to the interests or mission of the institution, whether it be a governmental agency, a corporation, a university, a hospital, a professional group, a municipality, a regional authority, or a labor union.[82]

The fact that government or the courts have been the preferred instruments is less significant than the general social expectations which they have been pushed to implement. Even with a political "sea change" such as took place in the 1980 election, it is not apparent that social expectations have changed. It is the instrumentalities of government, rather than the goals of government intervention, with which the various conflicting publics have become disenchanted. Thus new mechanisms are needed which are probably much more decentralized than federal or even state government, though exactly what they are can probably only be determined by experimentation. What seems to be needed is a means of compromising conflicting interests and goals at a much more "grass roots" level than is achieved at present.

Today people interact with many more other people than in the past, though many of these interactions may be more impersonal and partial. It may also be that the closest interactions occur with like-minded individuals, so that these interactions may only reinforce conflicting interests and goals rather than yield compromises which can be made effective in the political process. Businessmen talk only to other businessmen, academics to other academics, liberals to other liberals, conservatives to other conservatives, antinuclear activists to other nuclear critics. Supporting mythologies and pseudo-facts are frequently propagated within these closed circles, and provided with a legitimacy or credibility that has no basis in the real world.

New institutions and new forms of governance are needed which give effective representation to a wider range of public interests without compromising the basic social functions and viability of the institutions themselves. The problem is how to insure that any new governance mechanisms are accountable to the public as a whole and not just to particular constituencies whose perspectives they may be chosen to represent. One proposal is to have a much wider public representation and representation of affected interests on corporate boards or boards of nonprofit institutions with a significant public impact. Another is to have nonofficial community boards and advisory committees to represent all constituencies within a geographical area and to provide a mechanism for negotiating among those constituencies to arrive at coherent and concerted plans of action to advance the interests of the community as a whole.

There must also be ways of integrating consensuses arrived at in localized groups into a larger consensus extending over a wider geographical area or a wider set of economic interests. The problem at each level is to assure that the mutual gains that can be obtained from consensus-based actions outweigh the unilateral advantage available to any one group through manipulating the political or the judicial process. How to achieve this is not an easy matter, but one should not underestimate the importance of demonstration effects. Examples of successful experiments in consensus-building must be well publicized; indeed, one of the advantages of a decentralized approach is the potential it provides for small-scale experimentation and the possibility for rapid imitation of the successful experiments.

The transfer of experience from other industrialized countries to the United States can be another source for the introduction of social inventions and innovations. Although technology transfer of industrial technologies between countries is common and very successful, there has until recently been very little transfer of "social technologies," of practical experience in such areas as public transportation, urban waste management, control of the automobile in central cities, municipal finance, or community level energy conservation.[83]

The advantage of decentralized "grass roots" inventiveness in consensus building is that it provides for small nuclei of social invention and innovation that can, if successful, be replicated on a larger, government-supported scale once the possibility of success has been demonstrated on a scale that is usually small enough to be funded by private foundations or even voluntary initiatives. Here, indeed, is an area where the United States has a unique advantage because of its strong tradition of social invention and innovation through voluntary local initiatives independent of government.

Notes

1. From an account by Hajme Karatsu in "Quality Control: the Japanese Approach, the Better the Quality the Better the Cost?" pp. 19–24 in *Speaking of Japan*, published by Keizai Koho Center, Japanese Institute for Social and Economic Affairs, January 1981.
2. Anthony Smith, *Goodbye Gutenberg: The Newspaper Revolution of the 1980s* (New York: Oxford University Press, 1980), p. 23.
3. James B. Conant, *My Several Lives, Memoirs of a Social Inventor* (New York: Harper and Row, 1970).
4. Thomas P. Hughes, Chapter 2, this volume.
5. Warner R. Schilling, "Scientists, Foreign Policy, and Politics," in Robert Gilpin and Christopher Wright, eds., *Scientists and National Policy Making* (New York: Columbia University Press, 1964), cf. esp. pp. 155–157; cf. also C. P. Snow, *Science and Government* (Boston: Harvard University Press, 1961), esp. pp. 4–47.
6. *Scientific American*, Brochure for Prospective Advertiser, *The New Industrial Management*, 1966.
7. L. A. Sayles and M. K. Chandler, *Managing Large Systems* (New York: Harper and Row, 1971).
8. John G. Kemeny, "Saving American Democracy: The Lessons of Three Mile Island," *Technology Review*, June/July 1980 (Cambridge: M.I.T. Press); George M. Low, "Space Program Experience," in *Outlook for Nuclear Power* (Washington: National Academy of Sciences, 1980), pp. 32–38; T. H. Pigford, "The Management of Nuclear Safety: A Review of TMI after Two Years," *Nuclear News*, March 1981, pp. 41–48.
9. J. Servan-Schreiber, *The American Challenge* (New York: Athenaeum Press, 1968).
10. H. Brooks, "What's Happening to the U.S. Lead in Technology?" *Harvard Business Review* (May-June 1972), pp. 110–118.
11. Theodore Levitt, "The Gap Is Not Technological," *The Public Interest* no. 12 (Summer 1968): pp. 119–124.
12. E. F. Vogel, "The Challenge from Japan: Reaching a New Consensus on How to Deal with U.S. Competitive Problems," *Speaking of Japan* **2**, no. 1, (April 1981). (Statement before the U.S. Senate Finance Committee's Subcommittee on International Trade, Washington, D.C., Dec. 10, 1980).
13. The Franklin Institute, *Innovation and the American Economy*, the Fourth Franklin Conference, Nov. 1979, The Franklin Institute, Philadelphia, July 1980. Cf. esp. articles by A. M. Bueche and E. Mansfield; John H. Young and Scott Finer, *Technological*

Innovation and Health, Safety, and Environmental Regulation, draft report, Office of Technology Assessment 1980, Chapter VI; E. Mogee, *Technology and Trade: Some Indicators of the State of U.S. Industrial Innovation,* report to Subcommittee on Trade, Committee on Ways and Means, U.S. House of Representatives, by Experimental Technology Incentives Program (ETIP), National Bureau of Standards, USGPO, Washington, D.C., April 21, 1980; G. Christainsen, F. Gallop, and R. Havaman, *Environmental Health/ Safety Regulations and Economic Performance; An Assessment,* prepared for Office of Technology Assessment at request of Committee on Commerce, Science, and Transportation, U.S. Senate, August 1980; Office of Foreign Economic Research, Department of Labor, *Report of the President on U.S. Competitiveness, Together with the Study on U.S. Competitiveness,* transmitted to the Congress Sept. 1980, USGPO, Washington, D.C., 1981.

14. Jon Woronoff, *Japan: the Coming Economic Crisis* (Tokyo: Lotus Press, 1979), cf. Chapter 6, "Development Motors and Brakes"; C. Bose-Fischer, "Bonn Digs in Heels as Calls for Booster Programmes Become Louder," *The German Tribune,* no. 983, (April 5, 1981), p. 10.
15. Albert Rees, ch., Panel to Review Productivity Statistics, *Measurement and Interpretation of Productivity Statistics,* National Academy of Sciences, Washington, D.C., 1979. Cf. esp. Chapters 5, 6, and 9.
16. Dale W. Jorgenson, "Energy Prices and Productivity Growth," draft paper prepared for Harvard Energy Conference, Oct. 5, 1980.
17. Young and Finer, *Technological Innovation,* Chapter VI, "Regulation, Productivity, Growth, and Economic Performance"; Christainsen, Gallop, and Havaman, *Environmental Health.*
18. Young and Finer, Chapter VI, provide an excellent summary of all the recent literature.
19. See, however, H. Takano, "A Response to American Steelmakers, Clarifying Unfair Allegations," *Speaking of Japan* **2,** no. 1 (April 1981): 14.
20. Steve Lohr, "The Rise of American Exports," *The New York Times,* January 26, 1981.
21. Philip K. Verleger, Jr. "Alternatives Available to Reduce Impacts of a Supply Disruption on Transportation—Lessons from the Crises of the 70s," Kennedy School of Government, Energy and Environmental Policy Center, Discussion Paper E-80-10, December 1980.
22. R. A. Leone, W. J. Abernathy, S. P. Bradley, J. A. Hunker, *Regulation and Technological Innovation in the Automobile Industry,* draft report to Office of Technology Assessment, as revised June 2, 1980; William Tucker, "The Wreck of the Auto Industry," *Harper's Magazine,* November 1980, pp. 1–16.
23. Office of the Assistant Secretary for Policy and International Affairs, U.S. Department of Transportation, *The U.S. Automobile Industry 1980.* Report to the President from the Secretary of Transportation, January 1981, DOT-P-10-81-02; Lester R. Brown, Christopher Flavin, and Colin Norman, *Running on Empty: The Future of the Automobile in an Oil Short World* (New York: W. W. Norton and Company, 1979) Published for Worldwatch Institute.
24. OECD Secretariat, *Information Activities, Electronics and Communications Technologies—Impacts on Employment, Growth, and Trade,* OECD, DSTI/ICCP/80.10 (2nd revision), Vol. I; Colin Norman, *Microelectronics at Work: Productivity and Jobs in the World Economy,* Worldwatch Paper No. 39, Worldwatch Institute, Washington, D.C., October 1980.
25. Ichizo Yamauchi, "Japan's Electronics Industry Moves Ahead," *Economic Eye* **1,** no. 2 (Dec. 1980): 19–23, published by Keizai Koho Center, Japanese Institute for Social and Economic Affairs.

26. Howard Banks, "The Aeroplane-Makers," *The Economist*, August 30, 1980, pp. 5–24.
27. Interagency Task Force, *Private Sector Involvement in Civil Space Remote Sensing*, NASA, June 15, 1979.
28. Nicholas Wade, "France's All-Out Nuclear Program Takes Shape," *Science* **209**, (Aug. 1980): 884–889.
29. Steve Lohr, "Overhauling America's Business Management," *The New York Times Magazine*, January 4, 1981.
30. S. Kuznets, *Economic Change* (New York: W. W. Norton and Company, 1953); S. Hirsch, *Location of Industry and International Competitiveness*, (Oxford: Clarendon Press, 1967), esp. Chapter II.
31. J. Darmstadter, J. Dunkeley, J. Alterman, *How Industrial Societies Use Energy*, published for Resources for the Future. (Baltimore: Johns Hopkins University Press, 1977).
32. N. Bruce Hannay and Robert E. McGinn, "The Anatomy of Modern Technology: Prolegomenon to an Improved Public Policy for the Social Management of Technology," pp. 25–53 in *Modern Technology: Problem or Opportunity? Daedalus*, Winter 1980, issued as Vol. 102, no. 1, of the Proceedings of the American Academy of Arts and Sciences, p. 26.
33. Harvey Brooks, "Technology, Evolution, and Purpose," pp. 65–81 in *Modern Technology: Problem or Opportunity?, Daedalus*, Winter, 1980, cf. esp. pp. 65–66; cf. also Committee on Science and Public Policy, National Academy of Sciences, *Technology: Processes of Assessment and Choice*, report to the Committee on Science and Astronautics, U.S. House of Representatives, USGPO, Washington, July 1969, p. 16.
34. Theodore Levitt, "Management and the Post Industrial Society," pp. 69–103 in *The Public Interest*, no. 44 (Summer 1976), esp. pp. 85–89.
35. William H. Meckling, "Management of the Frequency Spectrum," *The Radio Spectrum, Its Use and Regulation*, proceedings of a conference in September 1967, *Washington University Law Quarterly*, no. 4 (1967) and no. 1 (1968), pp. 26–34.
36. Anthony Smith, *Goodbye Gutenberg: The Newspaper Revolution of the 1980s*. (New York: Oxford University Press, 1980). Cf. esp. Section II, "The Newspaper Industry in the United States."
37. Smith, Ibid., p. 73.
38. Smith, Ibid., p. 189.
39. Smith, Ibid., p. 96.
40. H. E. Meyer, "How We're Fixed for Strategic Minerals," *Fortune*, pp. 68–70, Feb. 9, 1981; G. Bylinsky, "One Answer to Imports: Wonder Iron," *Fortune* Feb. 9, 1981, p. 71.
41. Wickham Skinner, "The Impact of Changing Technology on the Working Environment," in C. Kerr and J. M. Rosow, eds., *Work in America: The Decade Ahead* (New York: Van Nostrand Reinhold/Work in American Institute Series, 1979), p. 208.
42. Karatsu, "Quality Control."
43. Melvin Kranzberg, Patrick Kelley *et al.*, *Technological Innovation: A Critical Review of Current Knowledge*, Volume I: The Ecology of Innovation, pp. 242–550, prepared for National Science Foundation, February 1975, esp. pp. 73–81.
44. Karen W. Arenson, "Housing After 50 Years, the Heyday is Over," *New York Times*, Business, Sunday March 29, 1981.
45. U.S. Department of Commerce, Office of Federal Statistical Policy, Bureau of the Census, *Social Indicators 1976*, December 1977, see Chapter 3, "Housing," esp. pp. 87, 88, 95, and 96.
46. R. S. Musgrave, "On Social Goods and Social Bads," Chapter 9 in Robin Marris, ed., *The Corporate Society* (London: Macmillan, 1974), esp. pp. 274–275.

47. Cf., for example, Edward J. Burger, Jr., *Science at the White House, A Political Liability.* (Baltimore: Johns Hopkins University Press, 1980), pp. 131–132.

48. Cf., for example, Ford Foundation, *The Local Initiatives Support Corporation: A Private Public Venture for Community and Neighborhood Revitalization*, May 1980; Office of Technology Assessment, *An Assessment of Technology for Local Development*, USGPO 1980.

49. Roger W. Sant, *The Least-Cost Energy Strategy: Minimizing Consumer Costs Through Competition,* including Technical Appendix, The Energy Productivity Center, Mellon Institute, Arlington, Va. 1979.

50. Cf., for example, Decision No. 92653, Jan. 28, 1981, on application no. 59537 of Pacific Gas and Electric Company, filed March 25, 1980.

51. C. K. Orski, "Urban Transportation: A Profile of the Future," *Vital Speeches of the Day* **46,** no. 1 (Oct. 15, 1979): 24–28.

52. Smith, *Goodbye Gutenberg*, pp. 247–254.

53. Smith, *Goodbye Gutenberg*, pp. 267–269.

54. Smith, *Goodbye Gutenberg*, pp. 262–267.

55. Smith, *Goodbye Gutenberg*, pp. 247–248.

56. Smith, *Goodbye Gutenberg*, pp. 274–278.

57. Jan Woronoff, Japan: *The Coming Economic Crisis*, esp. Chapter 6, "Development Motors and Brakes."

58. Lester C. Thurow, "Productivity: Japan Has a Better Way," *New York Times*, Business, Feb. 8, 1981.

59. N. Yashiro, "The Economic Rationality of Japanese-Style Employment Practices," *Economic Eye: A Quarterly Digest of Views from Japan* **2,** no. 1 (March 1981): 21–25.

60. E. T. Layton, *Revolt of the Engineers.* (Cleveland: The Press of Case Western Reserve University, 1971), Chapter 6, "Measuring the Unmeasurable, Scientific Management and Reform," pp. 134–153.

61. Wickham Skinner, "The Impact of Changing Technology on the Working Environment," p. 204.

62. Skinner, Ibid., p. 208; P. G. Herbst, *Socio-technical Design: Strategies in Multi-disciplinary Research.* (New York: Harper and Row Publishers, 1974), distributed by Barnes and Noble Import Division for Tavistock Publications.

63. Conference Board, "A Start Toward Better Working Relationships," a report of a conference sponsored by the Department of Commerce and the Department of State, Feb. 8–29, 1980; also M. Maccoby and K. Terzi, "What Happened to the Work Ethic?" in Joint Economic Committee, Special Study on Economic Change, Vol. I. *Human Resources and Demographics: Characteristics of People and Policy*, prepared for the special JEC Study on Economic Change, Dec. 12, 1980.

64. F. R. Anderson, A. V. Kneese, P. D. Reed, S. Taylor, R. B. Stevenson, *Environmental Improvement Through Economic Incentives*, Resources for the Future (Baltimore: Johns Hopkins University Press, 1977).

65. Federal Register, Dec. 11, 1979; R. Jeffrey Smith, "EPA and Industry Pursue Regulatory Options," *Science* **211** (Feb. 20, 1981): 796–798. Peter Nulty, "A Brave Experiment in Pollution Control," *Fortune*, Feb. 12, 1979, pp. 120–123; "Clean Air Dilemmas Stir Controversy," Conservation Foundation Letter, March 1981, Conservation Foundation, Washington, D.C.

66. *Comprehensive Environmental Response Compensation and Liability Act of 1980.* PL96–510, passed Dec. 11, 1980; J. Trauberman, "Superfund, a Legal Update," *Environment* **23** (March 1981): 25–29.

67. Conference Proceedings: *Utilities and Energy Efficiency, New Opportunities and Risks*, Oct. 23–24, 1980, Port Chester, N.Y., sponsored by New York Public Service Com-

mission, Economic Regulatory Administration, Department of Energy, Office of Utility Systems, pub. Jan 1981.

68. Peter W. Brown, ed., *Federal Legal Obstacles and Incentives to the Development of the Small Scale Hydroelectric Potential of the Nineteen Northeastern United States*, report to DOE by Energy Law Institute, Franklin Pierce Law Center, Concord, N.H., Jan. 30, 1979; Peter W. Brown, ed., *Fundamental Economic Issues in the Development of Small Scale Hydro*, Energy Law Institute, Franklin Pierce Law Center, Jan. 30, 1979.

69. Harvey Brooks, "Applied Socio-Economic Research in Energy and Environmental Policy," K. Arrow, C. C. Abt, S. J. Fitzsimmons, eds., *The United States and the Federal Republic of Germany Compared*. (Cambridge, Mass.: Abt Associates, Inc., 1979).

70. J. A. Casazzo, "The Engineer's Role in the Energy Crisis," *Public Utilities Fortnightly*, Feb. 16, 1978.

71. M. S. Crozier, S. P. Huntington, and J. Watanuki, *The Crisis of Democracy: Report on the Governability of Democracies to the Trilateral Commission*. (New York: New York University Press, 1975). Cf. especially Chapter V, Conclusions, pp. 157–171, and Excerpts of Remarks of Ralf Dahrendorf, pp. 188-195.

72. Cf., for example, "Nukes in the Marketplace," transcript of McNeil-Lehrer Report, Corporation for Public Broadcasting, Nov. 24, 1980; Frederic C. Olds, "Nuclear Energy in the United States: Outlook, Attitudes, and Implications," *Siemens Review* **47**, no. 4 (1980): 28–32.

73. See, for example, "The Clean Air Act—Under Heavy Siege," Conservation Foundation Newsletter, February 1981.

74. Brochure for The Health Effects Research Institute, Archibald Cox, chairman; Donald F. Kennedy, president, February 1981.

75. National Science Foundation, Joint Oceanographic Institutions, Inc., Santa Fe Engineering Services Company, *A Program for Scientific Ocean Drilling and Ocean Research in the 1980s*, Executive Summary, National Science Foundation, June 30, 1980.

76. Office of Technology Assessment, *Annual Report to the Congress for 1980*, Section V, "Organization and Operations," Congress of the United States, OTA, Washington, D.C., 1981.

77. John Busterud et al., *Nuclear Waste Management Process Review Forum, Final Report*, June 1980, sponsored by U.S. Department of Energy, U.S. Nuclear Regulatory Commission, Electric Power Research Institute.

78. L. J. Carter, "Sweetness and Light for Industry and Environmentalists on Coal," *Science* **199** (March 3, 1978): 958–959. Anderson et al., *Environmental Improvement*; NSF et al., *A Program for Scientific Ocean Drilling*.

79. R. E. Muller, *Revitalizing America: Politics for Prosperity*. (New York: Simon and Schuster, 1980), Chapter 8, "Quest for a New Vision: the 1980s," pp. 231–253.

80. W. Serrin, "Rohatyn, Going National, Doubts Free-Market Future," *New York Times* Metropolitan Report, April 21, 1981.

81. H. Brooks, "Technology, Evolution, and Purpose"; H. Brooks, "Technology Assessment as a Process," *International Social Science Journal*, UNESCO **25**, no. 3, (1973): 247–256, Paris.

82. H. Brooks, "Environmental Decision Making: Analysis and Values," L. H. Tribe, C. S. Schelling, J. Voss, eds., *When Values Conflict*, (Cambridge, Mass.: Ballinger, 1976), Chapter 6, pp. 137–152.

83. H. Brooks, "Politics for Technology Transfer and International Investment," R. Mayne, ed., *The New Atlantic Challenge*. (London and Townbridge: Charles Knight & Co., Ltd., 1975), Chapter 12, pp. 157–179; cf. also the series *Urban Innovation Aborad* and *Urban Transportation Abroad*, published by the Council for International Urban Liaison, Washington, D.C. since 1977.

chapter two

CONSERVATIVE AND RADICAL TECHNOLOGIES

Thomas P. Hughes

The most important conception in kinetics is that of "inertia." It is a matter of ordinary observation that different bodies acted on by the same force, or what is judged to be the same force, undergo different changes of velocity in equal times. In our ideal representation of natural phenomena this is allowed for by endowing each material particle with a suitable *mass* or *inertia-coefficient m*. The product *mu* of the mass into the velocity is called the *momentum* or (in Newton's phrase) the *quantity of motion*.

The Encyclopaedia Britannica
11th Edition

It arises from the circumstances that the number of institutional forms is restricted; from the fact that inertia and parsimony of spirit make us glad to employ established formulas; and likewise because, owing to the constancy of secular progress, it is difficult to recognize the moment when the choice of a new concept and name is desirable, when we should clear dead organisms out of the way, and when it would be well for us to introduce new outlooks.

Walther Rathenau
In Days to Come (1921)

From my perspective as a historian I see the United States now burdened by high inertia—the inertia of motion or momentum. Leveling off of production, falling productivity, and other indicators identify much discussed problems, but I see them as symptoms related to Newton's "quantity of motion," as well as problems. Although the quantity of motion of United States industry and technology is considerable, the country is experiencing small accelerations or changes in velocity or direction. I believe the historian is better able to perceive this inertia because he or she knows of the accelerations of nineteenth-century industrial America.

We are confronting a paradox. The nation now experiences technological innovation and related industrial development, but the innovation and the development increase inertia. To return to the mechanical analogy, the mass and the velocity are large, but rapid changes in velocity and, especially, shifts in direction are inconsequential. We are not an accelerating nation. Conventional wisdom believes that invention and development bring technological and eventually institutional and social change. The paradox is that American invention and development are now generally conservative in effect, thereby maintaining the direction and velocity of technology, institutions, and social structure.

This chapter will be about conservative and radical technologies. First, let me acknowledge the historians' limitations in addressing those active in contemporary affairs. In my opinion, historians are no better makers of policy than policy makers are historians. Of course there are exceptions, but they are rare. Historians, however, can contribute to policy making, and they should be taken seriously and listened to attentively. Historians can serve a useful function by asking questions of the past that are defined by the present. Furthermore, historians can search in the past for answers to questions or solutions to problems similar to those confronting us. Then the historians should ask the policy makers if the solutions of the past suggest courses of action in the present. A modest role, but an important one. It has been said that the person who does not know history will repeat its failures; similarly limited are those who do not know its successes. Perhaps I should add that because the solutions and the problems of the past are not identical with those of the present, there is ample room for imagination and invention.

Because technology is often mistakenly assumed to be neutral, its effects dependant upon how it is employed, we need definitions for the adjectives radical and conservative that will be used in this chapter. Conservative refers to technology, such as the means of production, that maintains the inertia of motion in the realm of production. Furthermore, conservative technology interacts supportively with existing organizations designed to nurture technology. Radical technology, in contrast, causes those sharp accelerations or changes in directions in production. Also radical technology requires new organizations or substantial changes in existing ones. Reference will also be made to radical and conservative inventors and innovators; they are the agents originating radical or conservative technology.

The concept conservative should also be qualified. The definitions given above simplify too much. Because I shall characterize American technology and its nurturing institutions as conservative and because I consider them open to moderate change, the concept of conservatism shades toward the conventional definition of liberal. It would be awkward, however, to refer frequently to liberal-conservative, so I shall simply use conservative and ask you to remember that it implies the inertia of motion with small accelerations.

Before developing the argument further, two difficult and complex concepts should be confronted: the kinds of messy ideas that the historian and the policy maker use. If the industrial growth of a country can be plotted over the long term—say, a decade—in a linear fashion, then we have an example of the inertia of motion. There are numerous perturbations in the curves of production and productivity, but if the second derivative is zero, then the phenomenon is one of high inertia.

What institutions nurture technology and production? The most obvious in this country are business corporations and government agencies. Also important are the research and development laboratories that are part of the corporation and agency structures. In addition, factories and mines are also the institutions of production. Less obvious, but important also are financial institutions such as banks and investment houses and political institutions such as legislatures and courts. Not to be overlooked are social organizations such as trade unions and industrial organizations. The list of institutions related to production and upon which it depends and exerts influence is endless. The point to be stressed here is that conservative technology does not disrupt the institutions nurturing it; radical technology radically transforms or even displaces them.

The argument developed in this chapter relates to the current concern with reindustrialization and revitalization. I suggest that our technology and institutions are conservative and that they interact in a mutually reinforcing and conservative way. The question then arises as to whether reindustrialization can be achieved without radical technology and concommitant radical changes in institutions, both government and private.

We should now turn to the past and ask such questions and raise doubts. My cases or examples from the past will be taken mainly from the history of rapidly industrializing or industrialized nations, especially Great Britain after 1750 and the United States and Germany after 1850. The case histories will be first about conservative technology and then radical.

The first example of conservative institutions and conservative technology is drawn from twentieth century German history.[1] The institution is *Badische Anilin-und Soda-Fabrik* (BASF), better known after World War I as I. G. Farben, of which BASF was a major constituent. Before and during World War I, BASF developed the Haber-Bosch process for the fixation through hydrogenation of nitrogen from the atmosphere to make ammonia. The process demanded high temperature, high pressure, and catalytic chemical technology. After the war, BASF, or I. G. Farben, had excess nitrogen production capacity, and, more fateful, the company had vested interest in hydrogenation in general. I refer to investment in scientists and engineers experienced in hydrogenation, workmen trained in operations, managers familiar with the related organization, as well as plant and equipment suitable for the high temperature, high pressure, catalytic processes. In theory, Farben could have turned to making automobiles, but with its vested interests the company prudently turned to other hydrogenation processes—first synthetic methanol and then synthetic gasoline. The subsequent history of I. G. Farben and these hydrogenation processes becomes increasingly interesting with the introduction today of synthetic oil processes, although the point to be made here is that these Farben moves manifest the inertia of technological momentum.

The scientists at Farben institutionalized their competence in hydrogenation through the establishment of the *Hochdruckversuche* (high pressure laboratory). Farben engineers, who had learned the technology in the hydrogenation of nitrogen turned in the mid-twenties to building the double-jacket reaction ovens, high pressure compressors, and corrosion-resistant pumps used earlier in ammonia synthesis and then needed for the mammoth synthetic gasoline plant at Leuna. Management organized and financed the largest research and development project (synthetic gasoline) in the company's history. The momentum was great and later swept I. G. Farben into a strange alliance with the National Socialists; but this is another episode.

There are also innumerable examples of conservative technological momentum in the history of American industry. The history of the Sperry Gyroscope Company provides examples. Elmer Sperry founded the Sperry Gyroscope Company in 1910 to institutionalize the further development and manufacture of his gyroscope inventions. Within several years he proved his gyrocompass, and widespread installation began. In 1922, Sperry's company astounded the world with the successful demonstration of the automatic ship pilot. Metal Mike took over from helmsmen ships, kept them on course for days, and understandably attracted attention as a remarkable robot. In fact, the Sperry Gyroscope Company was conservatively expanding a system the essence or core of which was the decade-old gyrocompass. The robot involved also complex feedback, expertise in which Sperry and his company had in abundance because of prior airplane stabilizer work. Feedback kept ships and airplanes on course against the reference provided by the compass.[2] Such expansion of systems through ingenious adaptation and invention by large organizations, private and government, is a widespread manifestation of momentum and conservatism.

The Sperry Company offers another example as we follow its move from searchlights into radar before World War II. This is a particularly enlightening example of momentum because superficially the continuity of searchlights and radar is not visible and also because the episode shows that what is taken on the surface to be radical technology is in essence conservative. Also, the Sperry example is a reminder that conservative technology involves invention and, especially, impressive development.

Elmer Sperry's first patented invention was a system of arc lighting involving, as arc lamps did in the 1880s, an automatic feedback control device for feeding and regulating the gap between the slowly consumed carbons. During World War I, the United States Army wanted improved searchlights that would find high-flying aircraft and dirigibles in the night sky. Sperry looked back three decades, drew upon his arc light experience, and developed a high-intensity searchlight incorporating not only his own ideas but the subsequent improvements of other inventors. The army found the high intensity searchlights introduced by Sperry and others the best detection device—and here the remarkable adumbration of radar—because "light energy seemed to be the only form of energy which could be projected through space rapidly enough to match the speed of the aircraft."[3] The Sperry Company went on to develop apparatus that calculated the velocity and direction of the aircraft by analyzing the movements of the remotely controlled searchlight that was detecting and following the object. Therefore, when the Varian brothers, Russell and Sigurd, along with William Hansen, invented the Klystron tube for transmitting microwaves, the Sperry Gyroscope Company bought these rights and continued along its line of detection work, now transmuted into radar.[4]

This account of radar has been greatly simplified and ignores other persons, companies, and agencies playing leading roles in radar history, but the Sperry part illustrates conservative momentum. Also, the episode shows how one must see beneath the surface of the events to underlying analogies in order to follow the line of development. It would be interesting to know if any of the inventors and developers associated with early radar were conscious of the analogy between detection with visible electromagnetic waves and invisible ones or were simply swept along by momentum.

The momentum extending from arclight to searchlight and then to radar was first maintained in the mind of the inventor Sperry and later in the context of the company. Generally, individuals such as Sperry display less inertia than companies and, therefore, their inventions lie less along clear lines of development. To state it differently, companies, agencies, and other large organizations with histories—including universities—are generally more conservative in research and development policy than individual inventors, but professional inventors also move along lines implied by their special competences and experiences.

In the nineteenth century, professional inventors other than Sperry also maintained momentum culminating in conservative technological change. Some inventors on occasions brought forth radical technology, but more on this when we turn to Alexander Graham Bell. First, however, we turn to the remarkably inventive conservative, Elisha Gray.[5] By 1875 Gray enjoyed an outstanding reputation as the inventor and developer of telegraph apparatus. He also founded a telegraph apparatus manufacturing firm, Gray & Barton, a successor of which is the Western Electric Company. By 1875 Gray knew, as did other inventors, that operating telegraph companies, such as Western Union, would pay handsomely for the solution to the problem of simultaneous transmission. The savings possible through reduction in metallic transmission lines or increase in density of use were obvious. Gray had the imagination to foresee that the harmonic telegraph, the principle of which was well known, might be the answer. In twentieth-century phraseology, he saw the possibility of transmitting simultaneously several messages by a composite wave form generated by tuning-fork transmitters of different frequencies and then unscrambling or filtering out the individual messages at the receiver.[6]

While developing a musical, or harmonic, telegraph for multiple transmission, Gray in 1874 accidentally observed a notable phenomenon: not only simple but composite tones complex enough for voice transmission could be transmitted electrically through a wire. About the same time in Boston, Massachusetts, the young professor of vocal physiology and elocution at Boston University, Alexander Graham Bell, was also seeking to solve the multiple-transmission problem with a harmonic telegraph. He, too, arrived at the conclusion that voice transmission was possible.[7] This is not the place to discuss the subsequent development of the telephone by both men; this is an opportunity to develop a thesis about momentum.

Gray was conservative; Bell, radical. Gray was the professional inventor; Bell, the amateur. With a long and successful record as an inventor of telegraph apparatus and with close contacts with the leading figures in the telegraph companies, Gray prudently avoided the distraction of the exciting telephone phenomenon. He energetically pursued the main chance and desultorily experimented with the fascinating possibility. As a professional who lived by his inventions, how could he do otherwise? He knew of no market and heard of no funding for voice transmission. Alexander Graham Bell was not so sensible, and he had his professorial income. Driven by an enthusiastic commitment, the amateur went ahead with the romantic notion. We know the rest of the story.

The lesson to be drawn from the Bell-Gray anecdote needs elaboration, but first another case of radical behavior. Eugene Jules Houdry, born in 1892 at Domont, near Paris, studied engineering and excelled in sports at the École des Arts et Métiers and won the Croix de Guerre in World War I. After the war he became intensely interested in automobile racing. This interest in dramatic, even romantic technology, led to Houdry's involvement in one early inventive endeavor. Bell's enthusiastic dedication to the idea of the telephone also suggests the more romantic side of invention. Houdry's fascination with racing cars also indicates a playful attitude to technology, as did Bell's experimentation in later life with tetrahedral kites and early airplanes. Playing with technology, as we shall see, is a characteristic of the radical inventor.

In France, Eugene Houdry in the 1920s became interested in two problems, one national and one more personal. He embarked upon the development of synthetic gasoline from lignite because France had lignite and little indigenous petroleum and upon the development of high octane gasoline because racing cars with increasingly high-compression engines needed it. The French government backed the synthetic gasoline project until rising costs, limited success, and the depression discouraged further participation. Houdry, however, had learned much about catalysts while working on synthetic fuel, and he saw the possibility of using catalytic agents to increase the yield of high quality gasoline from crude petroleum.

Persuaded that catalytic cracking was commercially feasible, he approached major oil companies, including Standard Oil of New Jersey. They were not persuaded, and Standard was already investing in the hydrogenation process of Farben as a potential means for increasing the yield of light fractions from crude. Not discouraged, Houdry sailed for the United States in 1930 to promote his catalytic process and persuaded the Vacuum Oil Company to take one-third of the stock of the Houdry Process Corporation, a development company. Houdry, with several engineers and scientists brought from France, experimented with a pilot plant in Paulsboro, New Jersey. The work was costly and because Socony-Vacuum did not supply sufficient funds, Houdry welcomed the support soon offered by the Sun Oil Company.

Because Houdry obtained his heavy steel reaction vessels from the Sun Dry Dock and Engineering Company across the Delaware, J. Howard Pew, president of the sister Sun Oil Company at Marcus Hook,

Pennsylvania, heard of Houdry's endeavors. The Sun Company was a small producer of gasoline, but had a reputation for the high quality (octane number) of its product. The owners, managers and engineers—especially Clarence H. Thayer among the engineers—also had reputations for flexibility and adoption of new technology and processes. They saw the Houdry product as a way of reinforcing their reputation for high-octane gasoline and agreed to take over one-third of the stock of the Houdry Development Company from Houdry's shares. Sun also supplied research and development funds and engineering experience. Since Houdry's equipment was so complex (some engineers said crude) the Sun expertise was valuable. Within several years, in 1937 the Sun Oil Company put Houdry catalytic crackers on stream.

John Enos, the historian of innovation in the petroleum industry, observed that Sun, a small firm, had a competitive advantage.

> The firm was not magnificent, not secure, but its management was willing to take the large risks inherent in the research, development, and installation of a *radically* [my emphasis] different process. Among the firms of the oil industry, the iconoclast is rarely the giant. And yet, if our calculations are correct, the iconoclast is certainly not the fool.[8]

Extending the analysis, I would stress that Houdry is a superb example of the relatively unburdened, radical, independent—even amateur—inventor. The Sun Company is an example of the organization with relatively low momentum. Also, flexibility was not simply a function of size; it was also a characteristic imbued in the organization by its risk-taking, innovation-prone owners and managers. "Sun was run by men whose position and personality were such that they were able to act quickly and decisively when it came to adopting a new process."[9] This judgment is reinforced by the fact that Sun supported Houdry after almost all of the major oil companies had rejected his project.

All of these examples, both conservative and radical, have been taken from private enterprise. The next instance of radical innovation is selected from the history of government enterprise, or more precisely, of mixed government and private enterprise. The episode involves one of the most closely integrated, momentum-prone institutions of modern history—the United States Navy. To inaugurate radical change in the navy is a prodigious feat, as historian Elting Morison has suggested in two masterful essays, one on the nineteenth and another on the early twentieth century navy.[10] I shall turn to a more recent example of the heroic naval innovation and innovator.

Unlike Houdry, Hyman G. Rickover was not an iconoclastic connoisseur of racing cars and the son of a steel magnate. Rickover arrived in

America by boat as an infant, the son of Polish Jews. Entering Annapolis and the navy from such a background did nothing to integrate him into the community then sometimes called "a floating plantation."[11]

Accounts are well known of Rickover's presiding over the development of the first nuclear-propelled submarine, the *Nautilus* and the first U.S. full-scale electricity-generating nuclear power plant at Shippingport, Pennsylvania. In these endeavors, he was a radical innovator, the introducer of radical technology. His achievements are all the more impressive because he introduced not only dramatic change in the navy, but within the structure of well-established American government institutions and industry as well. These prodigious feats stimulated many Rickover anecdotes, one having a Rickover aide accepting with equanimity for the Admiral a plot allocated in Arlington Cemetery despite a location in low, marshy ground. The aide anticipated that his chief would be out of there within three days.

Lord Zuckerman, in reviewing Richard Hewlett and Francis Duncan's book about Rickover's nuclear projects, insists that Rickover's style of innovation is unique.[12] On the other hand, I believe that one can generalize from his activities and find other instances of radical engineers whose strategies and tactics sustain the generalizations. Rickover, for example, finding high-inertia resistance in the navy and in the Atomic Energy Commission in the early fifties, "played the navy off against the Atomic Energy Commission and the Atomic Energy Commission against the navy."[13] Similarly, Rickover, trusted by the Joint Committee for Atomic Energy of the U.S. Congress, called upon this powerful instrument to cut through the maze of bureaucracy and overwhelm inertia. These tactics are not unique.

Calling upon outside influence to effect internal reform is a tactic used knowingly or through instinct by many innovators. Lieutenant William S. Sims, of the U.S. Navy, calling upon President Theodore Roosevelt to shake up the system and make room for a new technology of rapid gunfire is an outstanding instance.[14] Another military example has the well-known innovator, Winston Churchill, presiding over the introduction of radical technology during World War I. The British War Department investigated without enthusiasm tanks before World War I. After the imaginative Churchill was persuaded of the analogy between ships and armored cars, or "land ships," he ignored traditional areas of responsibility and supported the formation of the Land Ships Committee of the Admiralty, of which he was first Lord in 1915. Hoping to break the stalemate on the Western Front, he helped keep the project alive until the army used tanks at the Battle of the Somme in September, 1916.

Rickover acted in a great, but rarefied, entrepreneurial tradition when he invented social organization to nurture radical technology. His social inventions were unusually complex befitting the intricate array of established institutions upon which he had to lean and with which he had to contend: the Atomic Energy Commission, the United States Navy, and leading manufacturers such as the Westinghouse Electric Corporation and the General Electric Company. He also had to function within an environment in transition. In the late 1940s, under political and business pressure, the Atomic Energy Commission began to bring industry into nuclear reactor development. Furthermore, the tension between physicists and engineers for roles in reactor development—a tension extending back to the Manhattan project—was intensifying. In 1947 Rickover, then virtually unknown outside the navy, had the temerity to tell a meeting organized by the AEC and including Robert J. Oppenheimer that there were far too many physicists making decisions in the nuclear field. He wanted fewer committees and more engineers.

By 1949, after little more than a year of organizational activity, Rickover had organized engineers and many others. Under Rickover's prodding, a loosely structured *ad hoc* organization emerged involving the Navy Department and the Atomic Energy Commission, especially two relatively autonomous groups within those institutions: the Bureau of Ships and the division of reactor development of the AEC and the Westinghouse and General Electric companies. The purpose of the organization was the design and construction of a nuclear submarine. The responsibility for making it work was Rickover's. "He alone had created this strange alliance, and he alone could make it function."[15] It is not absurd to suggest that Rickover's making this creation work was more of an achievement than constructing a working submarine.

To summarize, let me cite the foregoing instances representing conservative and radical technologies. Badische Anilin-und Soda-Fabrik's nurturing of an extended series of high pressure, high temperature, catalytic processes; Sperry's development of a system of devices based upon gyroscopic principles and his company's move from visible light to microwave detection; and Gray's effort to take the next step in telegraph technology by multiplexing are our examples of conservative technology. Bell's breakthrough with the telephone, Houdry's determined push of catalytic cracking, and Rickover's relentless and resourceful cultivation of nuclear power serve to illustrate radical innovation and technology.

In conclusion, we should stress that the conservatives—BASF, Sperry, and Gray—did not encounter institutional resistance; the innovations were championed by substantial organizations. In contrast, Bell, Houdry, and Rickover were compelled to establish new organizations, deal with small and innovative companies, or foster *ad hoc* arrangements in the interstices of large organizations. These examples suggest that a hard line of definition between conservative and radical does not exist, but that there is a scale for differentiation based upon the measure of disruption the innovation causes in established institutions.

Radical innovations and technology cause disruption, stress, loss, and the likelihood of failure. The radical innovation forces a shift from the predictable to high-risk investment for the financier, from the manageable to disordered circumstances for the manager, from the controlled to unregulated and erratic process for the engineer, from the status of master to apprentice for the experienced worker, and often a shift from the workplace to the ranks of the unemployed for the unskilled. Momentum, or the inertia of motion, is disrupted by acceleration; a resistance by organizations and people who constitute them is understandable.

If in anticipation of radical effects, resistance is mounted, the innovation may never occur; if resistance is rearguard and futile, a legacy of resentment and intractability may follow. I should stress that reference is not to the alleged resistance of workers to labor-saving technology, consumers to new products, and regulators to innovation; the conserving inertia is the behavior of the highly organized institution defending the status quo—or a linear extension of it—by means of orderly system and routine.

If system and routine are some of the characteristics of inertia, what are the characteristics of radical innovation and technology? The styles of Bell, Houdry, Rickover, and other inventors and developers of radical technology reveal some. Curiosity, exploration, and play are one set of characteristics often found among the breakthrough inventors.

The concept of play denotes, among many other attitudes and actions, a "brisk movement or action, a space in which something can move," and "freedom for action, or scope for activity." The independent inventor takes the opportunity to play—Alexander Graham Bell is our example. Gray was burdened by the commitment to fulfill a defined need; Bell's enthusiasm was relatively purposeless in the economic or social sense. Another excellent example of the substance of play is the wisdom a few laboratory directors have shown in encouraging their most talented people to get out from under the burdens, the momentum of the laboratory, agency, or company and to play around with ideas.[16]

Another characteristic of radical technology is its crudeness as compared with the refinement of conservative technology. Warren K. Lewis, the MIT professor sometimes called the father of chemical engineering, told me that the large oil companies and their engineers found the engineering of Houdry crude and recommended rejection on those grounds. Since the untried is usually found wanting in the presence of the established and finely honed, I do not know whether the recommendation was naive or shrewdly calculated.

Not only is radical technology crude, but radical innovators characteristically offend our need for order and security. We are naive to assume that living inventors are nurtured and encouraged while they radically innovate. The historian celebrates the heroic American inventors, not their contemporaries. Elmer Sperry, who lived a long life as a professional inventor, was unable to forget and forgive early rejection and indifference even after he became rich and famous. He recalled "untold hardships" experienced because of "tradition" and "dogma," "those downright deterrents to progress and blights on the energy and will that bring progress."[17] I am told that after reading the biography of Sperry, Admiral Rickover remarked that Sperry had it comparatively easy.

Another characteristic of radical technology is unorthodoxy. It challenges the established organizational values of efficiency and expertise. An organization senses that it is likely to suffer a loss of highly valued efficiency and to commit embarrassing errors when forging into unknown areas. Productivity and production will decline during the troublesome period of transition. Let me illustrate this last point with a historical example. Statistics around 1890 pertaining to the growth of electricity production can be misleading if one is searching for long-range technological growth. The statistics often ignore the kind of electricity being generated. Hindsight reveals that if the increase was in the production of direct current, then the phenomenon was the lemming syndrome—the behavior of the mouselike rodents of the far north who move in a line in increasing numbers directly ahead and over a cliff. About 1890, the way to the future was not straight ahead, but a change of course into alternating current, which meant for a time a loss in momentum. The healthy, viable industry was the one changing pace, losing but changing ground, as it adjusted radically to changing circumstances and opportunities.

Finally we should ask how radical innovators and technology might be nurtured more effectively. First I shall suggest an institutional and then an attitudinal response. The case histories examined in this paper and the conclusions drawn from them suggests that cushions, couplers, and mergers are needed to lower institutional resistance. I am assuming that we will no longer tolerate the social miseries and human tragedies associated with the radical transformations of the British industrial revolution and with industrial urbanization in late nineteenth and early twentieth century America. If this is a reasonable assumption, then cushions, couplers, and mergers are needed to relieve the anxieties that foment implacable resistance to radical innovation.

By cushions, I mean modes of softening the hardships of change and loss. Support for reeducation and relocation are examples. These cushions, I must emphasize, are not the reactions we now employ to hardships, but the anxiety-reducing assurances that will lessen overt and covert resistance to change.

By coupler, I have reference to technological bridges that facilitate the transition from the old to the new. An excellent example in the technical realm is the devices invented by electrical engineers during the early twentieth century that allowed the old direct current system to be interconnected with the new and expanding alternating current ones. The increasingly obsolete was not destroyed, but allowed gently to fade away.

Mergers have an analogous function on the institutional level. Established institutions, both government and private, can join with the new and thereby have a stake in radical transformation. The institutional structure encompassing the old and new can provide a mode of moving capital and people into new regions without destroying the institutions and communities that sustain them. Again, I am not suggesting the couplers and mergers as reactions but as anticipations to disarm those made defensive and aggressive by the prospect of change.

The attitudinal response may be more difficult to achieve. A tolerance of the characteristics of the radical innovator and technology is needed. This means an openness to play in places of work, to crudeness in form and structure, to challenges to our sense of order and well being, and to skepticism about efficiency and expertise as the ultimate indicators of achievement. Tolerance of these characteristics also opens the door to the charlatan and the second-rate; no wonder that radical technology is so rarely risked.

Notes

1. The episode is described more fully in my "Technological Momentum in History: Hydrogenation in Germany 1898–1933," *Past & Present*, no. 44 (August 1969), pp. 106–32. On the history of hydrogenation and synthetic gasoline after 1932, see Wolfgang Birkenfeld, *Der Synthetische Treibstoff, 1933–1945* (Göttingen: Musterschmidt, 1964).
2. Thomas Parke Hughes, *Elmer Sperry: Inventor and Engineer* (Baltimore, Maryland: Johns Hopkins, 1971), pp. 277–83.
3. Hughes, *Elmer Sperry*, p. 221.
4. Edward L. Ginzton, "The $100 Idea," *IEEE Spectrum*, **12** (February 1975): 39.
5. For more on Gray and Bell, see David A. Hounshell, "Elisha Gray and the Telephone: On the Disadvantage of Being an Expert," *Technology and Culture* **XVI** (April 1975): 133–161.
6. Hounshell, "Elisha Gray," p. 144.

7. For the definitive account of Bell's invention, see Robert Bruce, *Bell* (Boston: Little, Brown, 1973).

8. John Lawrence Enos, *Petroleum Progress and Profits* (Cambridge, Mass.: MIT Press, 1962), p. 162.

9. Enos, *Petroleum,* p. 140.

10. Elting Morison, "Gunfire at Sea: A Case Study of Innovation," and "Men and Machinery," in *Men, Machines, and Modern Times* (Cambridge, Mass.: The MIT Press, 1966), pp. 17–44 and 98–122.

11. Jonathan Miller, "The Ancient Submariner," *The New Republic* **177** (November 12, 1977): 18.

12. Lord Zuckerman reviewing *Nuclear Navy, 1946–1962* by Richard G. Hewlett and Francis Duncan (Chicago: University of Chicago, 1974) in *Minerva* **XIV** (Spring, 1976): 118–26.

13. Zuckerman, *Minerva,* p. 123.

14. Morison, "Gunfire at Sea," *Men, Machines, and Modern Times.*

15. Hewlett and Anderson, *Nuclear Submarine,* p. 88.

16. Historian and metallurgist Cyril Stanley Smith has suggested the importance of play in invention. He mistakenly assumes, however, that the inventors of aesthetically pleasing objects and devices were not motivated and burdened by economic and social considerations. In many cases, the wealthy and the religious were their highly demanding patrons. C. S. Smith, "On Art, Invention, and Technology," *Technology Review* **78** (June 1976).

17. Hughes, *Elmer Sperry,* p. 294.

chapter three

INNOVATION AND THE GRANTS ECONOMY

Kenneth E. Boulding

Innovation is a deliberate change which somebody, presumably first the innovator, thinks is for the better. It is, therefore, clearly related to the perceived payoffs in society, especially for the innovator. A change which the innovator thinks is for the worse in his value estimates is very unlikely to be made. Perceived payoffs are primarily internal to the decision maker, which is what the economist is talking about when he develops the theory of "maximizing behavior." What is maximized, of course, is perceived net payoffs for the decision maker—usually called "utility." Utility is a function of the value system of the decision maker. This can be very complex. It does not, for instance, necessarily imply pure selfishness. There is nothing in the theory that prevents us from including benevolence and malevolence in it. For example, I may easily choose less of something if it will benefit people that I care for; this choice may have a larger utility for me than a malevolent alternative which will make me richer but the people I care for poorer. Consequently, if we are interested in why there is not enough innovation, or why there is innovation of the wrong kind or too much for a particular society to tolerate, we have to start with the perceived payoffs of potential innovators. In a society in which innovations are not rewarded in terms of the innovator's utility to that society, they will not be made.

An important aspect of *decision-making* which is particularly relevant to the problem of innovation is that perceived payoffs always include some element of risk or uncertainty. A decision is a choice among supposedly realistic images of the future, each related to a particular action on the part of the decision maker. Images of the future, however, always involve uncertainty, and the degree and the quality of the uncertainty is a very important aspect of decision making. And decision makers may vary a great deal in their fear of uncertainty. In estimating the present value or payoff associated with a particular image of the future, we tend to discount by time. Distant rewards count for less than more immediate ones. We also, however, discount for uncertainty. This is even more subjective than time discounting, and it is hard to put an exact figure on it. Nevertheless, measurement or no measurement, we do this all the time. If we are uncertainty avoiders, our decisions will tilt heavily to those futures which are more certain in our estimation, even though the potential rewards may not be as great as those uncertain futures which have high positive rewards, but also high negative ones.

Very little is known about what in life history and the learning process creates a person's uncertainty discounting capacity. There are people, like pathological gamblers, who actually have a positive preference for uncertainty, who purchase hope in the form of a lottery ticket or a bet, even though by any kind of measurable risk the present value of the expectation may be much less than what is paid for it. At the other end of the scale we have the pathologically cautious who never do anything, who only do today what they did yesterday for fear of doing something wrong.

Frank Knight[1] made an important distinction between risk and uncertainty. Risk is a property of an uncertain situation to which some known probability can be assigned. In writing a life insurance policy, because there is a large universe of similar cases, we can predict with fair accuracy how many people out of 10,000 will die before a given future date, unless, of course, there is a plague or a nuclear war. We cannot, however, predict whether a particular person is going to die in the next year, unless he is almost literally at death's door. Uncertainty, however, is a situation where we do not know even the probabilities for the realization of our image of the future. Innovators, on the whole, deal with uncertainty rather than with risk, although there is no reason why they should not take out insurance and things of that sort in the case where risk is known. The real innovator, however, must have a little bit of the mentality of the gambler; the hope of the gain must be much greater than the fear of loss.

The balancing of the hope of gain against the fear of loss is a very important element in decision making about innovation or investment. The greater the uncertainty, the more important this element becomes. Thus, a wheat speculator may be fairly sure that the price of wheat is going to rise by tomorrow. Normally, he will then invest in wheat up to the point at which the hope of gain, if his expectations come off, is balanced by the fear of loss if they do not. It is often the fear of loss rather than the hope of gain that actually determines the decision point. For the real entrepreneur and innovator, hope on the whole must be a more powerful motivator than fear, but increased uncertainty would still be a discouragement.

The perceived payoffs of any decision, and particularly a decision to innovate, are closely related to three factors in the environment of the decision maker. First, there is the production environment: What inputs does he think will produce what outputs and in what quantities? If these can be valued in some sense, then the larger the value of outputs per unit of the value of inputs—that is, the greater the value efficiency of the contemplated innovation—the more likely it is to be made.

Secondly, there is an exchange environment. Every person stands in a market environment where the assets he or she has, including money, can be exchanged for other assets at some rate of exchange or price. The relative prices of different things is a very important element in the value efficiency of production. The higher the prices of its outputs and the lower the prices of its inputs, the more likely is a particular production to be value efficient. The overall efficiency of any innovation or decision is closely related, therefore, to the relative price structure in the exchange environment, for this determines what is known as the "terms of trade," which might be thought of as exchange efficiency— that is, how much do we get in terms of some value per unit of what we give up in trade? As the relative price rises, this improves the terms of trade of the sellers and worsens the terms of trade of the buyers. The relative price structure obviously profoundly affects the terms of trade exchange relationships of all individuals participating in exchanges. A distinction can be made between actual terms of trade and actual transactions and potential terms of trade. Each individual has a potential exchange environment consisting of what exchanges can be made. This potential, however, may not be realized if there is ignorance, carelessness, or delay in the making of transactions.

The third part of the environment of a decision maker or a potential innovator is the grants environment. A grant is a one-way transfer of economic goods which effects the redistribution of the net worth of the parties involved. It is an accounting convention that exchange is of equal values, although, as Marx pointed out, this raises an interesting question as to where profits come from! If A gives B $100, however, and B gives A $100 worth of potatoes, their assets are rearranged but their net worths are unchanged. If A gives B $100 and B gives A just a nice smile, which the accountant does not recognize, there is a grant—A's net worth is $100 less and B's is $100 more. The grants economy, therefore, consists of that part of economic relationships in which net worth is redistributed. There are two parts to a grants economy: the first, direct grants, which may be, of course, either in money or in kind, in which net worth of the grantor is reduced and that of the recipient is increased by the value of the grant. The second is a very large area of indirect grants, redistributions which take place as the result of changes in the relative price structure, which redistributes net worth toward those who hold above average stocks of assets whose price is rising, and away from those who hold above average stocks of assets whose price is falling. Redistributions of net worth also take place as a result of a monopoly such as OPEC, which changes the relative price structure in favor of the monopolist and through innumerable different kinds of government regulations, quotas, licenses, and prohibitions which always change relative price structures in favor of some and to the disadvantage of others.

An interesting and quite difficult set of problems arises when the total net worth of society is increased. It is not easy to define "who gets the increase" as redistribution. A labor grant is an interesting example; volunteer services increase the net worth of some recipient but do not diminish the net worth of the grantor. Profit is another interesting example where the overall net worth of society is increased by the revaluation of assets above cost, for instance when they are sold. Here it is not always easy to define what is creation of new assets and what is redistribution.

The grants economy—that is, the total system of grants, both explicit and implicit, direct and indirect—has a profound effect on the overall dynamics of a society, mainly because it changes the structure of payoffs and hence changes decisions, which change subsequent production, consumption, accumulation, and may change beliefs and habits, voting behavior, or the support or non-support of institutions of all kinds. Social systems are also "ecosystems," which means also that they are "echo systems" in which any particular act may echo, re-echo, and reverberate about the system until the final consequences are very different from the initial change. The grants economy is no exception to this rule, and a grant, whether explicit or implicit, may have very different final consequences from what either the grantor or the grantee expect.

The grants economy can be divided roughly into the public and the private sector. The public grants economy, on the whole, consists of the tax system, tax payments, and negative taxes in the form of subsidies. For the individual, tax payments primarily fall pretty squarely in the definition of a grant. Certainly when I pay my taxes, my net worth diminishes while that of the government goes up. Taxes, when spent, presumably result in the production of public goods. These, however, do not usually get into private balance sheets, for the allocation of public goods to individual beneficiaries is an extraordinarily difficult, in fact, virtually insoluble, task. The private grants economy exists mainly in the family. About 30 percent of national income is redistributed within the family.[2] But it is also important in foundations and private charity, though this is not more than 1 or 2 percent of national income. Its qualitative importance may exceed its quantitative.

The two major motivations for grants I have described as love and fear.[3] Most grants are made from a mixture of both motivations. The grant to a bandit is made almost entirely out of fear. The taxes I pay, especially to the Federal government, are made at least 95 percent out of fear of the unpleasant consequences of not paying them and 5 percent out of identification with the purposes of the government. One can test this by asking how much one would subscribe to the government if it were financed by the United Fund. Even United Fund contributions are partly made in fear of what other people might say about you if you did not contribute. The large grants made to children in the family are often made out of identification with them and out of parental benevolence, but in many cases such grants have an aspect of deferred exchange, purchasing support in old age.

What, then, is the role of the grants economy in discouraging or stimulating innovation? It is clear that innovators themselves by taking risks may end up making large grants to the rest of society. This is particularly true of those whose innovations fail, as, in fact, most of them do. The motivations are clearly very mixed; there is undoubtedly in part a motivation for building up self-esteem, and there is also no doubt an important motivation in the hope of producing something that will make the innovator rich, by yielding a product the demand for which will provide the innovator with very favorable terms of trade. The motivation for self-esteem may also involve considerable elements of wanting to do good and to be a benefactor to the human race.

Governments frequently try to increase the motivation for innovation. One device is the patent law to give the innovator property rights to an invention. He is then permitted to take advantage of the exchange opportunities that these property rights imply. It is very hard to discover how significant the patent law really is, and how much innovation we would have without it. However, there is little doubt that it operates in the direction of encouraging innovation. All property rights are in a sense a grant from society, and the patent law is a particularly interesting example. It attempts to identify property rights in the innovation with the activity of the innovator, particularly, of course, of the first innovator. It is interesting that the property right in a patent is in a product, not merely an idea.

The copyright law is supposed to provide some property in ideas, or at least in the words that embody them, for their originator. The severe disapproval of plagiarism is an indication that most societies regard ideas as property, at least when expressed in terms of an original document. Plagiarism, however, is defined in terms of words, usually in written language. Oddly enough, a person whose ideas are stolen by others feels rather flattered by this. Here again, the motivation of self-esteem and contributing to the general welfare may be quite significant. A very important question which is of particular political significance at the moment is whether the state should create a grants economy for inventors, particularly inventors of theories, ideas, methodologies, and so on, which do not usually fall under the patent law. The National Science Foundation and the government laboratories, both social inventions, are examples of a grants economy, presumably based on the principle that new ideas and scientific discoveries are public goods which, once discovered, are the property of all and cannot be protected by anything like a patent law. In this case, therefore, direct subsidy is the best means of assuring an adequate supply of such public goods. In its early days, science was mainly supported by a private grants economy, from the rich or from endowed institutions like universities. Now it is moving increasingly into the public grants economy simply because of its increasing scale and expense, although private foundations still play a significant role. A very interesting question is whether government laboratories, like Los Alamos, have not actually perverted science toward human destruction.

The state, after all, is basically a threat system. It supports itself by threatening its own citizens into making them pay taxes or by extracting resources out of them by creating money, by which it can draw resources from the public by inflation. It is not surprising, therefore, that states, particularly sovereign states, devote a good deal of their resources to threats against foreigners, as well as against their own citizens. Innovation that is directly state supported, therefore, may be expected to go in large part to the threat system. The National Science Foundation is an exception to this because of a tradition that proposals for research in science should be judged largely by peers—that is, by fellow scientists. There is a certain danger in this that the subcultures of science may become too narrow and isolated from the general public and that hostility will develop between science and the rest of society, which could severely curtail the grants that society is willing to make and limit the rate of development of science itself.

Just where the public goods aspect of innovation and the private goods aspect actually meet or overlap is a quite difficult question. It is particularly difficult when it comes to the developmental aspects of innovation. There is a good deal in the remark, attributed, I think, to Edison, that an invention is 1 percent inspiration and 99 percent perspiration.

Certainly the translation of ideas into actual methods of production may involve much more human activity than having the ideas in the first place. Just how much development should be publicly supported and how much is satisfactorily dealt with in the private sector is a question to which there is certainly no easy answer.

In looking at the policy implications of these considerations, what is easy to overlook is that grants to society from the innovators may easily be more important in determining the rate of innovation and the success of innovation than grants from society to the innovators. A society which is infused with a spirit of ennui and disillusionment, in which the integrative aspects of the grants economy languishes because nobody really loves anything or anybody very much, may very well stagnate no matter what legislative incentives, tax remissions, and public and private grants are applied. This is perhaps the "supply side economics" of innovation. It has been much neglected in research on the problem, partly, no doubt, because it is very difficult to study. Just what creates a spirit of generosity, outgoingness, self-sacrifice, and pride in achievements of others is very little understood. It is certainly easy for a society to slide down into a kind of mean-minded, penny-pinching, ungrateful, self-centeredness that may be much more destructive to creative innovation than any defects in the patent system, the tax laws, or even government subsidies. One worries whether our own society is falling into this kind of threat to innovation that may be much more psychological than it is economic.

There may well be a critical level of benevolence in society above which it flourishes and below which it declines. Benevolence is a curious phenomenon, the study of which, again, has been much neglected. It does seem to have a certain quality of infection. We love because we have been loved. A very important aspect of society I call "serial reciprocity," in which B does things for C because A has done things for B.[4] This is quite likely to be a somewhat unstable system, and a few unfortunate experiences, some bad leadership, or perversion of the arts, may easily turn it downward. There are horrible examples of societies, like the Ik in East Africa, who seem to have descended into a virtual nightmare of mutual malevolence and distrust.

The role of trust in these processes is also of very great importance. Accountability is fine, but when it promotes mistrust it can be very costly. If we had a law, for instance, that everybody who dug a dry oil well was executed, we would not have much of an oil industry. One fears that public attitudes toward education and research, science and government, are constantly approaching this position. If we demand successes every time and are not willing to tolerate failure or even learn from it, we may get nothing at all. If we have too much trust, the cost of betrayal may be too high. But if we have too little, the cost of mistrust may be enormous. How to strike a balance in this matter is one of the most difficult questions in social policy, and certainly in social policy toward innovation and associated inventions.

Looking at these questions in another way, it could be argued that there is an optimum degree of disharmony, and that one can certainly have too much homogeneity in society. The great outburst of innovation in Britain in the eighteenth century, for instance, certainly had something to do with the fact that nonconformist sects were both strong and only mildly persecuted. The fact that nonconformists could not go to universities may well have contributed to their enormous contributions to technical and industrial innovation. There may even be something oddly beneficial in the perversity of having slightly persecuted minorities. Societies that have tolerated but not absorbed the Jews, for instance, have received an enormous grants economy from them in terms of innovation and creativity. Societies that have expelled and persecuted the Jews did so at great cost to themselves. Always, indeed, there seems to be an Aristotelian mean, and it seems remarkably difficult to find out where it is. Perhaps the greatest invention of all would be the one which would enable us to detect the subtle and imperceptible optima, through improving the processes of human learning, which now seem to be beset with so many pitfalls. Most assuredly the learning of beneficial social inventions and innovations will play an important, if not critical, part.

Notes

1. Frank H. Knight, *Risk, Uncertainty, and Profit*, 1st ed., 1921 (London: London School of Economics, 1946).
2. Nancy Baerwaldt and James N. Morgan, "Trends in Intra-Family Transfers," *Surveys of Consumers*, ed. Lewis Mandell, (Ann Arbor: Survey Research Center for Social Research, University of Michigan, 1973).
3. Kenneth E. Boulding, *A Preface to Grants Economics: The Economy of Love and Fear* (New York: Praeger, 1981).
4. Kenneth E. Boulding, *A Preface to Grants Economics*.

chapter four

THE INNOVATIVE MILIEU

Ralph Landau

INVENTION—ENTREPRENEURSHIP— INNOVATION

Definitions and Scope

Broadly speaking, what we are setting out to explore in this chapter is innovation, particularly technological innovation, in its various ramifications. Economists define an innovation as the first commercial application of a new or improved process or product.[1] Nowadays we would extend this definition to include a system such as the supermarket, time-shared computer, satellite communication, etc. In this context, there are, of course, also social innovations of the most important consequences, but I cannot possibly cover so broad a topic herein.

The innovative process consists of two distinct stages:

1. Conception or invention
2. Subsequent commercialization or exploitation

Thus the economist would point out that the invention by itself is not an economic good; it is only by commercialization that it becomes one.

Whereas the former task is that of the inventor, the latter role falls to the entrepreneur. Fundamentally, entrepreneurship is the process whereby people, money, markets, production facilities, and knowledge are brought together to create a commercial enterprise which did not exist before. Doing so does not have to embody an invention; it can simply be the founding of a new dry-cleaning establishment on a block which did not have one but where a need existed, or a new plant by a corporation using existing technology. If invention is involved, the importance of the entrepreneurial activity is even greater, because inventive novelty implies new risk[2] and completes an innovation.

Usually, the inventor and the entrepreneur are not the same person, though in rare cases they are. Thomas Edison fell into this classification, as did Henry Ford (a systems inventor!), Edwin Land, and Messrs. Hewlett and Packard.

There are various sub-classes of inventors or innovators. We may have an individual inventor, not supported by corporate facilities and staff, who makes inventions of fundamental importance and drives them forward by seeking support from various entrepreneurs or corporations. There is the corporate employee-inventor who has functioned within a corporate structure where he has had to champion his idea, probably in the face of much opposition at various times. There is also the purely technological entrepreneur who stimulates and participates to some degree in the research and development, but basically guides the creative invention of his colleagues and leads the group in the commercialization of the new technology, thus completing the innovation. As a Congressional Report[3] put it, "The technical leader is the individual who matches the world of science to the world of society, with a foot in management and a foot in science." I have had the honor of doing just that at Halcon for the 34 years of its existence.

In this chapter I shall be speaking of the innovative milieu in the more direct sense without addressing all the complex infrastructures, social systems, and historical evolutions which brought the societies in which I have operated to the point which I am describing. Thus, for example, I am not dealing with the educational system which produces so many of the technologists and entrepreneurs, although this is an extremely important part of the social inventions which have preceded the present period and which indeed have helped make the United States a unique milieu for technological innovation.

Kinds of Innovation

There are fundamentally two kinds of innovation:

1. The "breakthrough"
2. The "improvement"

This is much the same distinction that Professor Hughes makes in his chapter between "radical" and "conservative" technologies. Professor Nathan Rosenberg of Stanford, one of the few economists who specializes in technology and its effects, has elegantly described these two and their importance to the economy:

The growing interest in the diffusion of technology in recent years has functioned as a partial corrective to the heroic theory of invention. Inventions acquire their economic importance, obviously, only as a function of their introduction and widespread diffusion. . . . The central theme, on which I wish to elaborate, is that technological improvement not only enters the structure of the economy through the main entrance, as when it takes the highly visible form of major patentable technological breakthroughs, but that it also employs numerous and less visible side and rear entrances where its arrival is unobtrusive, unannounced, unobserved, and uncelebrated. It is the persistent failure to observe the rush of activity through these other entrances which accounts for much of the difficulty in achieving a closer historical linkage between technological history and the story of productivity growth.

He speaks first of complementarities, i.e., technologies seldom flourish in isolation,

It is a characteristic of a system that improvements in performance in one part are of limited significance without simultaneous improvements in other parts, just as the auditory benefits of a high-quality amplifier are lost when it is connected to a hi-fi set with a low-quality speaker. . . . This need for further innovations in complementary activities is an important reason why even apparently spectacular breakthroughs usually have only a gradually rising productivity curve flowing from them. Really major improvements in productivity therefore seldom flow from single technological innovations, however significant they may appear to be. But the combined effects of large numbers of improvements within a technological system may be immense. Moreover, there are internal pressures within such systems which serve to provide inducement mechanisms of a dynamic sort. One invention sharply raises the economic payoff to the introduction of another invention. The attention and effort of skilled engineering personnel are forcefully focused on specific problems by the shifting succession of bottlenecks which emerge as output expands.

He then logically treats of the cumulative impact of small improvements, saying,

a large portion of the total growth in productivity takes the form of a slow and often almost invisible accretion of individually small improvements in innovations. The difficulty in perception seems to be due to a variety of causes: to the small size of individual improvements; to a frequent preoccupation with what is *technologically* spectacular rather than *economically* significant; and to the inevitable, related difficulty which an outsider has in attempting to appreciate the significance of alterations within highly complex and elaborately differentiated technologies, especially when these alterations are, individually, not very large.[4]

Much of the literature is devoted to the "heroes" of innovation, whether inventors or entrepreneurs; very little is written about the improvements. I will discuss later why this overlooks some very fundamental aspects of the current milieu for innovation.

Figure 4.1 is an amusing cartoon which accentuates the point made here. I got this originally from Monte Throdahl of Monsanto.

Research & Development as an Aspect of Innovation

The real problem.   It is seldom appreciated that the invention, the R&D, usually costs far less than the process of first commercializing it. Where large capital-intensive projects are concerned, the R&D portion may be as little as 10 percent of the total cost; the remainder needed to convert the invention to an innovation may consume 90 percent of the total cost. Professor Mansfield and his colleagues at the University of Pennsylvania have been studying this kind of relationship and, as might be expected, found in seventeen chemical innovations that R&D costs ranged from 7 percent to 71 percent of total project cost.[5] Though for new products the percentage is higher than for new processes, the important conclusion is that the invention itself (the R&D expenditure) usually costs less than half of the amount spent on the innovation. The realization of this relationship has been slow in coming to the majority of economists, who pay little attention to the economics of technology, and it has been even more tardily perceived by politicians who listen to the economists, despite their mixed advice. Although R&D expenditures, as a percent of GNP, have been falling, it is only recently, as we in the United States start to probe into the reasons for our apparent decline in innovation, that it is being discovered that it is the risk-taking entrepreneurial side of the innovation process, causing the greater part of the innovation, which has been faltering.[6]

Changes are occurring.   It is not surprising that this should be the case; most politicians take it for granted that the recent past (presumably satisfactory) state of the nation will continue forever unchanged. In our country, the great successes and dominance of our technology since World War II have blinded us to the changes which are taking and have taken place both internally and externally. Alexis de Tocqueville, writing nearly 150 years ago in his extraordinary *Democracy in America*, put his finger on one of the reasons for this apparent blindness,

> In America there are no nobles or literary men, and the people are apt to mistrust the wealthy; lawyers consequently form the highest political class. . . . They have therefore nothing to gain by innovation, which adds

Fig. 4.1 Identity and role—aspects of risk taking and innovation.

a conservative interest to their natural taste for public order. . . . These (legal counselors) secretly oppose their aristocratic propensities to the nation's democratic instincts, their superstitious attachment to what is old to its love of novelty, their narrow views to its immense designs, and their habitual procrastination to its ardent impatience.

Doesn't that have a contemporary ring to it?

Of course, a democracy must be founded on a legal system; we cannot, alas, do without lawyers. Allied with them are the accounting, financial, and other professions which control the pace of social and political change. Opposed to this social conservatism, which our legally trained politicians prefer, stands the technologist, the architect of *change*.

Recent trends have been putting even greater control of our business institutions into such conservative hands, in parallel with their already overwhelming control of the political institutions.[7]

Value of Technological Innovation to the Economy

This constant and increasing tension is all the more surprising if we look at the underlying reason for the great growth in the American standard of living which has occurred since de Tocqueville wrote. Thus, if one estimates that the average GNP growth in real terms over the last decades has been perhaps 3.5 percent per year, technology alone has contributed somewhere between 25 and 50 percent of this growth. The effect of technology on productivity growth is treated as a "residual" by economists after calculating labor and capital factors.[8] Notwithstanding the general inability of econometricians to measure this factor with precision, it is clear that technology advances are the key element of healthy sustainable growth.[9] Dr. Richard Atkinson, former head of the National Science Foundation, has spoken of 45 percent.[10] These numbers usually understate the facts, because the *quality* of new products and of new capital investment cannot readily be measured, and these have also been increasing. But clearly, technology has been a major contributor to the advances in the American standard of living since the middle of the last century. Dr. William Nordhaus,[11] a former member of the Council of Economic Advisers, says, "the effect of technological change substantially outweighs that of increase in capital stock." The full import of this finding has not yet been digested by many; it is laden with policy implications.

In a critical survey made in 1978 by Professor John M. Logsdon of George Washington University, an extensive review of the economic studies of R&D and innovation was undertaken. He stressed that most economists would agree that the body of findings from the studies reviewed confirm the intuitive notion that technological change is *the* driving force behind economic growth, and has been for the United States the major source of such growth and of increases in productivity for a very long time.

Technology, therefore, lies at the heart of economic growth, which is a vital necessity since without it our free society and solutions to our economic and social problems are doomed. No one can hope to freeze the status quo without dictatorship. A declining economy, needless to say, is unthinkable in a democracy. It has furthermore been clearly noted that political freedom exists only in free-enterprise countries where political and economic power are separated, which are also countries that lead in technology, although all capitalist countries are not politically free.[12] The present chapter has implicit throughout it my firm conviction that our Western industrial world and especially the United States cannot retain their freedom without growth, and that such growth must, as in the past, be based on technology.

The Uniqueness of the American Climate for Innovation in the Past

In general, history shows that the significant effects of socio-economic trends take years to emerge. Britain began to decline as an industrial power as far back as the nineteenth century when technological leadership passed to the United States and Germany, but the full effects are only now visible and are going to be difficult to reverse. Thus, we must be grateful to our predecessors who created in this country a climate, a milieu, for innovation, including especially technological innovation that has led to the growth described in the preceding paragraph. The essential features of this climate are the coupling of the invention to the market, which is the unique role of the entrepreneur seeking his fortune (whether he be an individual or a corporation), and a lack of pervasive influence by government or extraneous controls. Perhaps two illustrations from our own experience, which describe our earlier research activities and their consequences (more or less successful) will assist in understanding the processes that have historically been occurring for over a hundred years. They both occurred at the tail end of this long period and at a time when conditions in the United States and abroad were still favorable for the entrepreneur, who had available a huge surge of technological developments accumulating from the war era.

Ethylene Oxide and Glycol. Our concept in starting our company was to develop enough of our own proprietary technology so that we could find customers for it—a market. It was possible to organize and run a new company in those days with minimum capital and no worries about governmental report requirements. Indeed, it wasn't until a year or so after incorporation that we inadvertently discovered that we were not legally qualified to do business in New York, but it was not serious!

Our first effort commenced in 1946, when I started to think about ethylene oxide (a very widely used chemical intermediate) by direct oxidation. Indeed, during my preceding work at the M. W. Kellogg Company, we had played around with fluidized bed catalysis for such a reaction, but subsequently at Scientific Design (as our company was originally called), we realized that the nature of the kinetics and the difficulty of scale-up made this version a very long shot indeed, which we could not afford. We opted for fixed bed studies at our inception, and naturally the research on catalysts occupied a central role.

My own interest was immediately attracted to the newly available spherical supports, and I reasoned that these would lower the pressure drop and thereby save energy. Some of my colleagues disagreed, saying that this very saving in pressure drop would reduce heat transfer in the catalyst and cause uncontrollable hotspotting. Such doubts can only be resolved by experiment, unfortunately requiring prototype tubes and conditions which could not be built in our recently purchased (by stock) laboratory on 32nd Street off Park Avenue in New York. I therefore concluded an agreement with the English firm of Petrochemicals Ltd., for whom we actually had done some initial consulting work, to erect and operate such a pilot plant in exchange for an exclusive license in the United Kingdom. Thus, by sheer necessity we seemed launched on the path to licensing this first ''baby'' of ours. The pilot unit was started up in 1948 and worked quite well. Of course, both research and engineering activities were proceeding simultaneously in New York. We did indeed prove that a spherical catalyst was an advance over the art, and so I had the great satisfaction of ultimately receiving my first patent of importance.[13]

But now the Jekyll-and-Hyde part of our philosophy appeared. My partner, Harry Rehnberg, and I began to ask ourselves why, with this kind of patentable potentiality, we should not seek to exploit the process by our participation in manufacture, at least in the United States. So in 1948 we started a series of discussions that led us within a year to the concept of a plant to be owned equally by Sears Roebuck, Shell Chemical, and ourselves. Sears would contribute a significant market, Shell had excess ethylene at Deer Park, Texas, (this was before the advent of merchant ethylene, which did not appear until the 1950s), and we had a rather unique technology. (Although Union Carbide was known to have a direct oxidation plant and process, its nature and results were carefully guarded secrets, and the process was clearly not available to others who were using the old chlorhydrin process.) Thus, in this technology we were not dealing with a "breakthrough" invention but with a market opportunity for which we were seeking to develop a competitive position. We almost pulled it off: Sears approved it, and Shell seemed favorably disposed. But somewhere in that giant worldwide organization, resistance developed, and the project collapsed.

However, we must have stimulated Shell into doing their own research work in California, because a few years later they announced a process of their own was available for licensing and began an intense effort to find their first commercial customer. At the same time Shell (UK) bought Petrochemicals Ltd. at Carrington, England, acquiring our previously granted exclusive license on ethylene oxide along with it. Clearly, this could be a source of embarrassment to us both. We had several meetings in an attempt to resolve the matter, and, if possible, to join forces so that they could utilize our exclusive license in the U.K. Our version used air and theirs used oxygen, and we claimed that one or the other might be preferable in a particular situation. In addition, by 1953, we could point to our first commercial plant at Naphtachimic in Southern France (Shell's first such plant came quite a few years later in the United States). But none of these conversations were successful, and we have competed universally ever since. Subsequently, we developed our own version of an oxygen process as oxygen prices came down, and thus we became the only organization to offer a customer either version as his situation required.[14]

As a result, we have between us built most of the free world capacity, except for Union Carbide's plants. Ours account for 30 companies with 66 plants in 24 countries, approximately one third of the total. We have been responsible for most of the Communist countries' buildup of this key organic intermediate, with appropriate licenses from the U.S. Government (recently interrupted by the Afghanistan crisis, which for the first time interfered directly with our international activities of 33 years' standing).

This evolution obviously made it impossible for us ever to revert back to an exclusive manufacturing position, except in one country, Brazil, where we have a minority interest in Oxiteno S.A. Shell, much later, did build several plants of their own using their technology, but they have, as a result of the early decision to license by both of us, a relatively small share of the market. Both of us have probably gotten rather small royalties over the years, and we both spend substantial amounts in continuing research to improve the processes, primarily for the benefit of our licensees. A gracious tribute to us by the leader of this vigorous competitor was recently published.[15]

So we learned some valuable lessons: the world is the market for industrial companies; an exclusive technological lead may last a short time; competition may develop rapidly and internationally; and, once the cat is out of the bag, obtaining a strong manufacturing position or market share becomes difficult if not impossible. I have written a paper on the many ways in which technology can be dissipated.[16] We also learned how expensive complete technology transfer can be relative to the more or less standard royalties prevailing in our industry.[17] Licensees prefer to pay only for your research successes and leave the failures to you!

Terephthalic Acid. Early in the 1950s, I was visited by a friend who explained the market situation in terephthalic acid manufacture. This principal component of polyester fibers (originally discovered by Rex Whinfield at the Calico Printers Association in England around 1940, who sold out to Imperial Chemical Industries) was being made by nitric acid oxidation of p-xylene. The latter was a commercial curiosity in itself, and the first manufacturer (Standard Oil of California) had an exclusive supply deal with DuPont, the exclusive licensee for the fibers in the United States. My friend had just bought the rights to the new Imhausen-Witten process from West Germany, which made dimethyl terephthalate (an equivalent component of polyester fiber) by a four-step process using air oxidation of p-xylene. Unfortunately, there was no p-xylene to be had. Could we come up with a substitute feed, in the para position?

I have told the story of this development elsewhere,[18] but, in brief, the idea occurred to my colleagues and me that p-diisopropylbenzene might be that feedstock. We had built a cumene plant for Allied Chemical using aluminum chloride as the alkylating agent and were familiar with the fact that di- and tri-isopropyl benzenes were formed as by-products and could be recycled to extinction. What about p-isomer? If it could be isolated, the m- and o-isomers could also be recycled. We set up an experimental unit, and, sure enough, found we could distill the p-diisopropylbenzene away from the other two isomers. We had a process and ultimately a patent.[19]

Unfortunately, by this time our friend had found a source of p-xylene, and his interest had evaporated. So I decided to look for a way to oxidize p-diisopropylbenzene with air directly to terephthalic acid and produce a process which might be licensed in competition with Imhausen-Witten, particularly abroad where p-xylene supplies were nonexistent (in fact, our first licensee was Pechiney in France). We found that certain catalysts like manganese salts worked very well, and our research seemed on the verge of success: a cheap feedstock and an air oxidation process! Then we were thrown into an interference with Shell on our oxidation patent applications, which we eventually won. We also issued some other patent cases in this area.[20] During this rather intensive legal battle, I became curious as to what the affect of the anion in the catalytic metal salts would be. We discovered that bromide ions had a strongly accelerating effect on the oxidation, but the yields remained the same. Here is where a brilliant research chemist is indispensable. Drs. Saffer and Barker followed this lead by trying the bromide-metal catalyst on p-cymene, which has one methyl and one propyl group attached to the benzene ring, with excellent results, and then it was but a step to p-xylene (two methyl groups on benzene). The first experiment succeeded beyond our wildest hopes. Methyl groups were preferentially oxidized with the aid of the bromine and thus we had discovered the air oxidation of p-xylene to terephthalic acid the Mid-Century Process, as we called it.[21]

We knew we had a "biggie," as Johnny Carson would say, a real "breakthrough," and there was no doubt we would try to get into the manufacture ourselves, instead of licensing it. In this instance also, DuPont seemed to be our natural partner or customer, but it didn't work. Quite by accident, we heard that Standard Oil of Indiana was interested in p-xylene manufacture, so we tried out the idea of a joint venture on them. It seemed at the time like the mating of an elephant and a mouse, and instead Standard offered to buy us out altogether. We succumbed for money (our first real personal gains) and a licensing agency abroad. We were, as might be expected from our earlier experiences, quite adept at this and succeeded in finding numerous licensees in Europe and Japan. In fact, the first plant ever to use this technology in the world was built by Mitsui Petrochemical Industries in Japan! But Standard would not license in the United States until very recently (to Du Pont), opting instead to build its first commercial plant by oxidizing mixed xylenes and separating the acids by a self-developed process. They have gone far since then, including the development of the Pure Terephthalic Acid Process as a direct feed for fiber manufacture. They are the unquestioned leaders in terephthalic acid technology, but in Europe and the Far East they have encountered significant competition from their previous licensees. Their policy changed over the years, but a license once granted is difficult to revoke!

The moral here is also clear: valuable new technology *is* capital.

From these true stories, in which I was an active technical participant as well as functioning chief executive, let me underline some significant features of those earlier days which illustrate the general nature of the great period of American industrialization and technological domination:

1. No governmental planning or restrictions of significance were present, either here in the United States or abroad. The "invisible hand" of Adam Smith was thus free to act, and the seeking of private gain by so many individuals and companies in many countries, including Halcon-SD and myself, led to the development of the great synthetic polyester fiber (Dacron, Fortrel, Terylene, etc.), whose properties and cost were made so attractive that market penetration was tremendous. No one could have planned these results. The same was happening in other fields of technology in which we were active, although I have not described them here.

2. We were able to move freely across national boundaries and felt no negative impact from our government, if not any positive help (except for Marshall Plan aid in the early 1950s). Washintgon simply did not figure in our activities, and I doubt that I visited that city more than once or twice throughout our first 15 years (which is the time period embraced by the two case histories described). If that were only true today! The most optimistic description of the relationship of government to business today would be "neutral," but in many instances it is inherently adversarial.

3. The time to get things done was much shorter then—at the most two years to complete a plant project abroad and 18 months in the United States. This is important in holding down all costs, including R&D, thereby helping to accelerate the pace of technological change.

4. As a result of the war devastation and the need to rebuild, the receptivity for new technology and for risk-taking was greater abroad than in the United States. Already in our country, the first signs of the increasing conservatism and other trends to be described later were appearing. We circumvented them by moving abroad first (as in the first case history) or by selling *exclusive* rights to a very valuable technology as an inducement (as in the second case history).

5. We developed an ethylene oxide "club" as a feature of our international licensing activities which I believe was a first in our industry. We arranged meetings at frequent intervals, rotated among the plants of our licensees, which I chaired, and where both formal papers and informal discussions took place. In this way, and by other exchanges, we ensured that the improvements obtained by our operating licensees were made available to others in the club (under appropriate contractual arrangements). Over the years our process got better and better as a result, and the licensees all benefited as well. In later years, I believe, Shell emulated our example, at least for a while. This might be deemed an innovative "system."

6. Capital costs of new plants, and indeed their scale, were much

lower during these years, so that entry costs by various companies were much lower than today, with attendant reduced risk; vigorous competition ensued.

The Historic Source of Major Innovations: The Individual or the "Outsider"

It has been implicit in some of my preceding remarks that individuals have played a disproportionately large role in the creation of geometric (breakthrough or radical) change since the Industrial Revolution. Names such as Colt, Morse, McCormick, Bell, Edison, Ford, Sperry, and others of the nineteenth century are very familiar as inventors and innovators. Indeed, most of the great multinational corporations of today began with a creative process inspired and promulgated by an entrepreneur.

Mr. William J. Casey, a prominent former SEC (Securities and Exchange Commission) Commissioner as well as Export-Import Bank head, has described this phenomenon in a broader context using more literary language:

> From the time of Pericles through Elizabeth I down to Polaroid, the cutting edge of dynamic societies has been the innovator, risking his own savings and those of others having confidence in him, whether on the waves of the high seas and new horizons or those of high technology and new services. Almost every new technology that has given a life to the American economy has come from a new company, struggling in a garage or venturing out to obtain needed capital from the public. For example, railroad car manufacturers did not pioneer the development of automobiles; neither did established automobile companies develop the first airplanes. The most dramatic recent example of this phenomenon has been the development of the semi-conductor industry. Again, the original innovations came from new companies, and not from the manufacturers of vacuum tubes. In the past fifteen years, the industry has passed through four relatively distinct generations of technology, and each successive phase has been led by a new entrepreneurial enterprise. Today, innovative young companies, once established, are not only growing faster, but actually creating more new jobs and tax revenues than the giants of American industry.[22] [This is confirmed in a recent study at MIT by Professor Birch.]

To this list could be added the electronic computer (Mauchly and Echert's ENIAC at the University of Pennsylvania), Chester Carlson's Xerox process, Sir Frank Whittle and the jet engine, catalytic oil refining, the helicopter, the Polaroid camera, and in our own industry, I have already mentioned polyester fiber. Furthermore, these should include Ziegler-Natta chemistry, the single most important *post*-World War II discovery, resulting in a rare Nobel Prize for industrial innovation. Professor Ziegler was a German independent researcher in a small laboratory, and Professor Natta taught at an Italian university.

What these examples and those of Mr. Casey are intended to emphasize is that it is the "outsider" who is more often the key "breakthrough" innovator in our society. They use their own money for risk-taking or persuade others to advance it to them on faith, all with the lure of making money on a *big scale* if successful. Frequently, too, it is the successors to these men who make the real profits, but so does society as a whole as the technology diffuses more widely. That has been the driving force for innovation in this country. As I mentioned in our own case histories, we discovered early on that as "outsiders" we would face resistance from the American chemical industry, and this has always been true. My remarks should not be construed as bias against innovation by individuals versus large or small organizations. They can and do all innovate, albeit perhaps differently from case to case. There are certainly many large very innovative companies, and indeed, some innovations can only be handled by large organizations with enormous resources in capital and research.

Our results, nevertheless, have confirmed that the "outsider" position is one of real importance. In some recent publications,[23] I have presented tables of the major commercial "breakthrough" processing developments in the chemical process industry over the past 25 years. The total number I found was 31, out of which 9 came from us; since then we have added a tenth. Most large companies produced only one or two major discoveries during the last quarter-century, and only a few had as many as four. It is also interesting to see that more than half of these were developed by nonchemical "outsider" companies, e.g., by oil companies, contractors, and research firms like ourselves.

A recent example of the profound influence an outsider can have on a whole industry was the introduction into the United States of the radial tire by Michelin of France, which doubled the life of the typical automobile tire from 20,000 to 40,000 miles. It destroyed the entire economic foundations of our tire industry, among other effects, very nearly wrecking Firestone, the number two company in that industry.[24]

Another example is in the American Telephone and Telegraph Company, whose Bell Telephone Laboratories have been responsible for some basic "breakthroughs" which the company itself, because of its huge internal market could justify commercializing and hence could take large developmental risks.

II. The Current Milieu for the Individual Innovator

The Last Ten Years

The symptoms of declining innovation.   Over the last ten years, a national consensus has slowly been building that the great wave of innovation described above was coming to an end or at least slowing down markedly when compared with our international competitors, such as West Germany and Japan. The innovation study conducted at the direct request of President Carter and initiated by Dr. Frank Press focused national attention on many of the symptoms of this decline, such as:

1. Sharply decreasing number of new technology ventures (and indeed of new companies of any kind) as the venture capital markets dried up
2. Declining export competitiveness
3. Declining share of GNP which goes to R&D
4. Declining savings as a percent of GNP, and reduced capital formation

It is not necessary herein to go over the same ground, as it has been widely published and discussed. Dr. Press has given an excellent summary of the results at the earlier conference on creativity, invention and technology, and it is greatly to his credit that he did indeed consult the innovators themselves in approaching the problems and their solutions. In essence, this consensus recognizes that innovation and productivity are the national issues of the 1980s, and the importance of tax and regulatory policy changes to improve the climate is widely recognized. He foresees a new era of government-industry-labor cooperation to these ends with the recognition that all parties will not only have to maintain such an improved economic climate, but will have to do so more steadily and with less of a near-term orientation for "quick fixes" or instant results.

The changing basic climate.   When it comes to accounting for this decline, opinions vary widely. It is not, in my opinion, any malevolent force that has brought this about, but basically an over-confidence bred by a seemingly endless cornucopia of abundance. This over-confidence led us to assume that all kinds of social innovations, some good, some bad, were possible, but in their totality they were beyond the capacity of the system to sustain or correct itself (in this respect, perhaps we are seeing another manifestation of the Kondratieff long wave of economic-social history).

My own experience points clearly to the current trend toward egalitarianism, to the redistribution of wealth, as a significant contributor to the present situation. People are either innovators or imitators; wealth creators or spenders. The prevailing moral philosophy of the Western World favors the imitators, the less productive, the "disadvantaged," the spenders. Innovators, the wealth creators, are a small minority, and in a democracy cannot politically defend themselves readily against such an onslaught. Yet we have never been in greater need of innovation, particularly because the trends cited above are not susceptible to the accountant/lawyer standard "quick fixes"—fiscal and monetary manipulation, wage and price controls, restrictions on the movement of goods, services and capital across international boundaries, etc.

New wealth formation has been greatly inhibited by the trends of the last 20 years or so. Why is this so important to society? Because old wealth tends to be conservative and non-risk-taking; new wealth tends to support new ventures, growth stocks of the riskier types, and new technology investment. A healthier stock market would be a reflection of such underlying changes. A good example of this distinction may be found in a recent history of the richest family in America, the Mellons.[25] In their heyday, as investment bankers, the Mellons supported many new entrepreneurial companies, and their income today mostly comes from these earlier investments in Gulf, Alcoa, Mellon National Bank, Koppers, Carborundum, First Boston, General Reinsurance, Hewlett-Packard, etc. But today they are much less active in this type of funding. As one review of the book said,[26] "today the Mellons are no longer accumulators, they're only spenders." Undoubtedly an exaggeration, but the book does make my point clear.

The Barriers to Innovation Today

As Gene Bylinsky of *Fortune* magazine, who studies the technological entrepreneur, has said, "Because the best of the breed [entrepreneurs] is good at clearing hurdles does not, of course, mean that the hurdles—such as burdensome regulations and punitive tax schedules—are good for society."[27] Entrepreneurship is fragile, and in the best of circumstances requires Tender Loving Care from society. What are these hurdles, these barriers? One can write books on these themes, and I will do no more than outline some of the well-known facts (I have written extensively elsewhere on these subjects):[28]

Inflation. Unquestionably, the United States is currently suffering from the worst sustained siege of inflation since the Constitution. What is more, all the industrialized democracies are caught in it, to a greater or lesser degree.

Inflation brings high interest rates (anathema to the new venturer and the equity investor), a dearth of risk capital as a result of the general decline in capital formation, and the increasing domination of a short-term view in all aspects of economic endeavor. The increasing institutionalization of the stock market as a result of the relative unattractiveness of equities to the individual investor in inflationary times results in less risk taking investments, and the ERISA rules only abet this trend.

Inflation is a tax on the productive members of society (both past and present) to support the unproductive members. As such, it is another manifestation of the same basic change in climate alluded to above. Whatever the individual causes or contributors to inflation may be, the cumulative effects have become deadly in their impact on innovation. Indeed, inflation at current levels, if long continued, would destroy our private freedoms, especially all the countervailing agencies and institutions, leaving us naked before the all-powerful state, which alone has the power to print money.

In a very recent article,[29] Edwin Mansfield has stated in summary:

> R&D, through its effects on the rate of productivity increase, can significantly restrain the rate of inflation in the medium and long run. High rates of inflation damage the workings of the price system and impair the efficiency of practically all economic activities, including R&D. Findings suggest that the percentage increase between 1969 and 1979, in total real R&D expenditures, has been exaggerated due to the inadequacy of the gross national product deflator as applied to R&D. He adds, "Serious inflation tends to discourage investment, including investment in certain kinds of R&D, because it increases uncertainties concerning relative prices in the future."

Inflation can never be arrested if some elements of our society, by virtue of economic or political power, succeed in mitigating its effects on themselves at the expense of others, as by indexing of pensions, wage increases with COLAS aggregating far above productivity increases, escalating farm values, taxation ratcheting up faster than incomes, increasing unemployment while wages of the employed rise higher, etc. What this process has accomplished is a very uneven sharing of the burden or "tax" of inflation, but in the end it has been a tax on productivity, capital investment, innovation, and job creation.

Uncertainty. Perhaps above all else, business (large and small, but particularly the entrepreneur) needs a higher degree of certainty by way of general economic and legal climate than we have seen over recent years, e.g., price and wage controls, changing energy rules, changing tax laws (usually for the worse), increasing opportunities for time-consuming (and often unfounded) litigation, overlapping and frequently contradictory regulatory rulings by different federal agencies, and changing accounting principles. The only exception I can see to this need for certainty is inflation, where the problem is too much certainty— growing feelings that inflation (but with an increasing intensity) is here to stay. As George Will has said, "It is said business is reluctant to invest because of 'uncertainty.' Actually, business reluctance reflects the virtual certainty that inflation will remain intolerably high and that government will require corporations to devote more resources to environmental and other social purposes."[30]

No democratic government can or should try to iron out all of the bumps in the economic road. There will be periods that are better than others, and that's a risk that has to be taken. But a free-enterprise democratic government does have the responsibility of not moving in fits and starts, by applying short-term fixes to long-term problems, changing direction like a broken field runner. The government should confine itself primarily to the macroeconomic sector and the correct policies to aid the supply side of the economy, and leave the detailed decisions— the fine tuning—ranging from such minutiae as OSHA's design for safe lavatories (now mostly rescinded) to wage and price controls for thousands of firms (abandoned not long ago) to the pluralistic wisdom of the market and the individual enterprises.

An insightful analysis of the secular and cyclical changes which have taken place in the investment climate is given in a recent study by a prominent Wall Street firm.[31] They point out that in recent years investors in stocks have come to demand a higher risk premium over bond yields, citing such factors as inflation, the strains on the international lending institutions and on trade, the rising tax burden on the productive sector of the economy, the decline in the quality of earnings and assets brought about by the rapid buildup in unfunded corporate pension fund liabilities, the problems of the Wall Street firms themselves, etc. Thus, investors today are emphasizing risk and return, rather than return exclusively. And if the risk premium is unusually great for investment in large companies, then it must become astronomically high for new risky enterprises, and this is why so few can "make it" or even get started today.

The former Chairman of the Council of Economic Advisers, Alan

Greenspan, has also written about investment risk assessment by business today.[32] He says, "Thus, the critical focus of economic policy in the western world has got to be to reduce these abnormally high risk premiums. They have created a private decision-making atmosphere which gives short shrift to long-term benefits and costs and undue emphasis to the short run." He stresses that because most Western governments have been activist in policy and will not reduce such intervention overnight, it is all the more important to lower taxes on business and capital. These cuts, he says, are not a "permanent substitute for lowering risk, but to the extent that after-tax returns to capital are increased, they will offset some of the high-risk (discount) in the capital investment process. . . . There is no substitute for a non-inflationary environment if prosperity is our goal." Another expression of this viewpoint was also contained in the 1977 *Economic Report of the President* (p. 28).

I have been and am involved with many such investment decisions both as a shareholder and as a chief executive officer, and I can only confirm the truth spoken by these authorities on the subject. Entrepreneurial risks require a longer time horizon than is currently demanded by investment and uncertain conditions today—about a four-year span, which accounts for the currently low price/earnings multiples of many even blue-chip stocks. In other words, the short-term view of managements entrusted with the public's money is no accident.

In a speech in Vienna[33] I amplified some of these subjects as they affect the international chemical industry investment patterns. Another example from current industrial real-life situations may be found in the aluminum industry. It is no secret that new technology is within reach to permit utilization of the abundant domestic clay resources instead of imported bauxite, which therefore would contribute greatly both to national security and the balance of payments. But the inflationary bias of our economy is reflected most acutely in the rapidly escalating costs of building new and risky capital projects of this kind. In addition, the long range policy of the United States with regard to the structure of power costs and pricing, choice of fuels, environmental restrictions, forced recycling, etc., is undecided if not contradictory. Any such conversion of the aluminum industry to domestic raw materials requires not only adequate profitability expectations such as a closer approach to replacement pricing (taking the competition from other materials into account) but a reduction in the uncertainty levels so that longer range earnings need not be so heavily discounted as at present. There are examples like this throughout the United States, in old as well as new enterprises.

Taxation.   The previous sections on inflation and uncertainty have explicitly referred to tax policy as one of the critical areas that needs corrective action. The basic recent struggle in this area was the matter of capital gains tax rates, which were raised in 1969 and further raised in 1976 to a level essentially equal to earned income taxation (i.e., 50 percent). Before 1969, the capital gains tax was of course 25 percent.

Immediately after the 1969 Revenue Act started this further effort at egalitarianism or "social justice," venture capital began to dry up, and so did the new technology equities market. From 1,298 new equity issues in 1969, the number had fallen to about 50 in 1977. A bitter struggle ensued in 1978, when President Carter actually proposed to raise the capital gains taxes even further in the alleged interest of fairness, but a Congressional revolt developed, accurately reflecting feelings in the country. And in the Revenue Act of 1978, President Carter very reluctantly accepted a reduction in the capital gains tax to 28 percent. In fact, although the Treasury adamantly predicted this reduction would lose the government several billion dollars in revenue, it now appears that capital gains revenues have actually *increased* by hundreds of millions of dollars, as a result of increased economic activity. This gives support to the argument that there is some substance to supply side economics.

What this fight reflected was the realization that not only were such high tax rates stifling venture capital and the entrepreneur, they were really confiscating capital in an inflationary era, since many of the so-called "gains" were due solely to inflation. Obviously, this was likewise unfair to innovators and risk takers, but society was the loser. Is 28 percent "fair" today? No. It is still among the highest in the world.[34] It needs to be indexed with length of time the security is held, but there is no doubt that this reduction, the only significant pro-capital formation reform in a decade, partially revived the equities market, despite the persistence of inflation, and the number of new equity issues has greatly increased to about 200.

Currently, a great deal of attention has been given both by the Carter administration, the Senate Finance Committee, and the Republican Party to providing some form of relief on depreciation allowances, which is of particular concern to corporations. In an inflationary era, currently allowable depreciation is based on historic cost rather than replacement cost, but replacement cost will be much higher as a result of inflation and technological obsolescence. Hence, if we are to gain increased innovation, at a minimum of the nonheroic type, adequately rapid cash flows must be obtained by the corporation in the early years to justify its initial investment.

Regulation. Society must have some regulation and there always has been. For example, businessmen have never been allowed to shoot the competition's chief executive or burn down his building. Where a society concludes a particular minimum of social behavior is necessary to the marketplace, the rules must be binding on all so that no one competitor can have an unfair advantage. Thus, there exists a legitimate basis for regulations as to child labor, pollution, sales to potential enemies, unsafe factories, toxic or otherwise unsafe substances or products, and the like.

However, it is essential that regulation-makers come to understand that every regulation has its price and its practical limits—in the costs people pay for the goods and services produced, in competitive posture worldwide, in impact on jobs and in possibly stifling new investments that, if successful, can mean a better quality of life for our people. In other words, there are always tradeoffs and each must be carefully weighed, debated, and decided.

There just must be some way that the businessman and those financing him can have reasonable assurances in advance of investing thousands or more likely millions or billions in a product, process, or plant that he won't go broke after proceeding in good faith, because the rules of the game change in the sixth inning. The *speed* with which the ground rules have been changing in the last decade has had a great deal to do with the declining growth of the economy, which for private plant and equipment, excluding pollution control expenditures, was an average 4.3 percent per year in 1965–70, 3.3 percent in 1970–75, and declined further to about 2.5 percent per year in 1975–77.[35] Dr. Charles L. Schultze, former Chairman of the Council of Economic Advisers, has written a very interesting analysis of some of these problems[36] and favors economic incentives over regulations wherever possible.

An entrepreneurial company often finds that premature "going public" soon alters its innovative attitudes, and the management, under SEC and other external pressures, shifts to a short-term and less risky strategy, often to the detriment of its long-term growth and innovation. Privacy is a great help to boldness, but if boldness is to be sustained over longer periods of time, the investors must be confident that ultimately they will be rewarded by financial gains,[37] as mentioned above.

Paradoxically, technology is the key factor in improving the environmental and safety aspects of our society. In our industry, and indeed directly in our own discoveries, more efficient processes are also the ones that pollute the least and are the most energy-efficient, and as described above, some of these replaced toxic oxidants with either air or oxygen.

Generally, it is the older industrial establishments that have the greatest environmental and energy problems, and the solution to these (often a very costly one, as in steel) also requires more capital formation and higher technology. But here we also see the contradictory effects of different regulations and policies as inhibitors of progress. The current clean air "offset" requirements that "old pollution" be reduced before new plants can be built in the area (except within a narrow "bubble") means that "old polluters" have been granted a high value by the law, under a sort of grandfather clause, whereas the new, efficient, less polluting plants (such as those I mentioned above) cannot easily be built, except in remote locations far from existing infrastructure such as modern industry demands. This further penalizes the economies of our larger cities and industrialized areas, where many jobs have already been lost. Since these new plants also require much more capital than the depreciated older plants, it is difficult for business or investors to justify so long-term a risk as these new technologies represent. With energy, the environmental regulations have a different but no less stultifying effect: substitution of "clean oil" by "dirty coal," which the nation's economy requires, is retarded.

All of these points have been thoroughly analyzed in a colloquium report of the National Academy of Engineering[38] which also reflects the realization of how we have come to favor the passive acceptance of less investment in favor of more consumption and redistribution.

The Ever-Continuing Availability of Would-Be Entrepreneurs in the United States Is Still a "Plus"

Despite these barriers and recent developments, as Gene Bylinsky said, and as I have often observed, there is no lack of entrepreneurial desire on the part of Americans. Quite the contrary—only the hurdles are more formidable.

Science carried an article[39] on the reasons for the failure of West Germany and Britain to encourage growth of new companies based on technological innovation, yet these two countries were leaders in this activity in the nineteenth century! There is no permanent advantage for any country, unless it is assiduously cultivated. No better contrast to our historic aptitudes exists in this regard than the experience of the Soviet Union, which outlaws entrepreneurship and decentralization in favor of central planning. An example is found in *The Technological Level of Soviet Industry*.[40] In the chapter on the chemical industry, the author's conclusion is,

Compared with most Western countries, the Soviet research effort and total output of scientific papers are probably considerable, but the overall quality is such that it does not appear to have made a proportionate impact on world science. Also, the Soviet research effort does not seem to have generated any really important and original innovation, which could be successfully scaled up to mass production.

I have personal experience of some of these matters since we are doing several projects in the Soviet Union, and I think the author has hit the nail pretty well on the head. There is no pattern of innovation in Russian industry simply because the penalties of failure far outweigh the rewards of success. It is much safer to buy complete plants and the financing that goes with them from the West! On the other hand, they have very able chemists who do a lot of good work.

The Current Milieu for the Corporate Innovator

Much of what has been said in this chapter about the changing milieu for the individual innovator is equally significant to the corporate innovator. There are, however, some features that are particularly pertinent to the latter.

The Power of the Learning Curve and its Significance to Innovation.

Dr. Edwin A. Gee, who spent many years at the senior management level of DuPont, wrote in his book[41] about that company's experience with the enormous benefit of small accumulations of knowhow or innovation: "If a rule-of-thumb generalization is drawn from the nine cases [cited in Chapter 8] it would be that 50 percent reduction in mill cost in constant dollars is an attainable goal in five years from the point of regularized operation, with a further 50 percent reduction (75 percent overall) attainable in a further ten years (15 years overall)." He points out that this is usually accomplished by both economy of increasing scale and economy of the learning curve, i.e., progressive technological improvements.

Mr. Robert Malpas[42] has discussed the application of learning curves to the chemical industry and been given some experience by Imperial Chemical Industries where he served as an executive director until becoming president of Halcon International.

The power of such a cumulative learning curve is not only beneficial to the firm (if it can retain possession of the knowhow long enough), and justifies much R&D (probably most): it also explains why the real technological breakthrough (which is usually patentable) becomes more difficult the longer the competing product or technology has been in commercial operation. Figure 4.2 shows a typical learning curve, in which the cost of production in cents per pound is plotted against the cumulative production. (This kind of logarithmic plot has been shown to be valid for an astonishing number of goods, from automobiles to plastics to commodity chemicals, when plotted in constant or real dollars. It should be remembered that such a curve ought reallly to apply to a fixed or standard product, but as mentioned before, much R&D is devoted to improving quality, so that a simple curve cannot alone express all the benefits to the firm of cumulative small improvements.)

Let us suppose that at point C a real technological breakthrough oc-

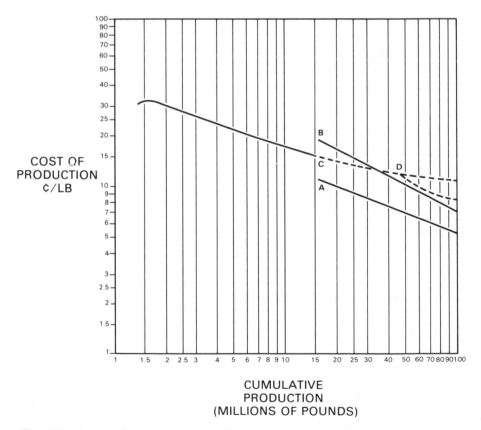

Fig. 4.2. A typical learning curve reflecting cost of production and cumulative production.

curs, and a new curve could be commenced from Point A. Clearly, the new technology will be expected to replace the old, as its learning curve accumulates experience. However, life is usually not that simple for managements: the new technology may involve an initial cost at B *higher* than C. Nevertheless, if it follows its expected slope, its learning curve should eventually cross over below the curve where C would be if continued, and sooner or later B will take over. But it is a real act of courage and faith for a company to commit to an investment for B when the same company has plants at C, and with even a chance that a further minor technological breakthrough such as D might improve the slope of the learning curve so that B might hardly or never catch up!

The current worldwide inflationary trend also means that capital investments required for new "world scale" plants are now enormous, being usually $100 million or more (often, much more) so that every incentive exists to "ream out" existing plants by debottlenecking or other short-term improvements. Furthermore, in such a high-inflationary era, the worst inflation comes in the capital cost of new plants, as compared with the consumer price index. In addition, the time to build such plants has substantially lengthened, so that the time cycle mentioned by Dr. Gee is probably considerably longer now. This combination results in expected returns, especially for complete or grass-roots plants, based on a commonly used method such as the discounted cash flow (where the time value of money is important), which look increasingly threadbare, compared, for example, with the advantages of buying an existing company. Only the small incremental improvements would continue to show satisfactory discounted cash flow rates of return under these conditions, and this explains why for many companies a major part of their capital budgets consists of such improvement projects costing just a relatively few million dollars each. Other project are often mandated energy- or pollution-related. These conditions also account for the current perception by many in the congress and the business community that the proposed Capital Cost Recovery Act of 1979 (providing for accelerated depreciation) is one very important step to improve the discounted cash flow returns for new and larger projects. High inflation has also meant low price-earnings multiples on Wall Street, and acquisition fever consequently runs high.

A good example of this situation may be seen in our company's innovative activity with the Atlantic Richfield Company (ARCO) through the Oxirane Group in the field of propylene oxide. I will present a brief account to illustrate some further points:

Propylene Oxide. As a result of our successes in both vapor and liquid phase oxidations, such as I have described for ethylene oxide and terephthalic acid, I came to the obvious conclusion that we ought to find a better method than the chlorhydrin process for making propylene oxide. After all, ethylene oxide had once been made by the chlorhydrin technology also! But the chlorhydrin process had been invented around World War I and was obviously very far down its learning curve, so we knew we had our work cut out for us.

Why were we so interested in better ways of making propylene oxide? Because it is the major component of urethane polymers, engineering materials of extraordinary versatility and flexibility for a huge variety of applications. Discovered by the Bayer Company in 1937, the industry has grown to its present stature because of the independent market-oriented competitive work by many organizations: DuPont, Upjohn, and ARCO in isocyanates; Wyandotte and C. C. Price in polyethers; Carbide, Mobay, and Houdry in one-shot urethane foam technology; and many others. But, curiously, everyone was still using the chlorhydrin process for propylene oxide! What an opportunity as well as a challenge!

We soon eliminated all the simple approaches; our silver ethylene oxide catalyst just burned the propylene to CO_2! Much laboratory work was expended, and eventually the chemists brought me a partial oxidation noncatalytic process which they felt would show good economics, even if not too selective. Similar work was being done in Europe by Naphtachimie and others. I tried to arrange some kind of collaborative research effort, but the more we talked the more convinced I became that this was the wrong path, and that we needed a much more selective process in order to compete with the well-established chlorhydrin process. We thought of liquid-phase oxidation because of milder conditions, possibly leading to high yields. The successful outcome was first described at the World Petroleum Congress in 1967[43] and subsequently in a recent article in *Chemtech*[44]: basically we had discovered the catalytic liquid-phase epoxidation of olefines with hydroperoxides to produce olefine oxides. We had another "biggie!"

But there was never any doubt in my mind that this technology would be our ticket to chemical manufacturing. It took us three years to find a partner,[45] and the apparently improbable name of ARCO became the most logical of all candidates, for several reasons:

1. They had been following a similar direction of research to ours.
2. They had a compatible feedstock position.
3. They were a significant customer for some of the products.
4. They were also interested in entering into chemical manufacturing (they had little in the way of chemical operations at the time) and in providing substantial financial backing.

Thus was born the Oxirane venture, which became one of the great success stories of the petrochemical industry. Only very recently, in June of 1980, we sold our half to ARCO for 270 million dollars and other considerations.

It is again noteworthy that it was two "outsiders," ARCO and Halcon, who entered this field of technology against the establishment (Dow Chemical). In so taking the plunge, which required plenty of "guts" in view of all I have said above, we nevertheless knew that history shows in time that really good technology will choke off the old, and the stand-pats lose position; however, their managements of the day probably have already retired! Such breakthroughs and risks are only justifiable if the patent protection can give a reasonable head start. And, looking backward, if investments continue to be made in the new technology and its learning curve improves, it becomes much harder, particularly in inflationary times, for new entrants to come in later with the highest capital and operating cost plants which require full depreciation and with no possibility of averaging return on historic or partially amortized capital, unless they, in turn, make a new technological breakthrough and take the big risks!

The Significance of the DCF in an Era of Inflation Militating Against Breakthroughs

Some 20 years ago, the process industries generally adopted use of the discounted cash flow method (DCF) to evaluate the desirability of new investments, a system not widely employed by the individual entrepreneur or the venture capitalist. This method, which calculates the present worth of the future stream of cash to be generated by the new investment, based on the then prevailing interest rates for money (which of course were much lower than they are today) was used to discriminate among investments, e.g., if an oil company had to choose from among a pipeline, a refinery, a chemical plant, or an oil well, and capital was as usual limiting, the higher the DCF the more likely it would be that a particular project would be allocated scarce capital funds. This system was favored as a way of expressing the responsiblity corporate management felt in their handling of other people's money.

As inflation developed and interest rates rose for the reasons mentioned above, the expected DCFs for many projects went down, or, conversely, the future potential earnings in depreciating dollars would have to be enormous to justify an investment in today's currency. Thus, fewer risk-taking projects can be justified, and innovation is reduced, except for the small plant improvement type which typically still commands a satisfactory DCF. Thus, in one large company with which I am familiar, nearly half of the very substantial annual capital budget is devoted to small improvement projects of the order of 3 to 10 million dollars. It perhaps should be no great surprise that this firm's equally substantial R&D budget is over 95 percent devoted to improvement studies, with the small remainder truly categorizable as entrepreneurial or risk-taking. As a result, there are few truly innovative capital projects going forward, except those directly connected with maintaining its main lines of business on an appropriately diversified basis.

Further Response of the Corporation to the Changing Environment

A recent speech by a successful chief executive officer, J. Paul Lyet of the Sperry Rand Corp., which is itself quite entrepreneurial, is revealing:

> I certainly have no illusions about the fact that public attitudes today require both the entrepreneur and the CEO to be manager, politician, diplomat, journalist, and public figure. There is no clear line between what is directly and indirectly related to the corporate bottom line. Also, what is true today for the largest corporations will become true for the smallest. The entrepreneur of the eighties, like the CEO of today, will find his time frame too small to cover all the internal and external demands of him.
>
> This broadened scope means that the corporate staff has to pick up a share of the external burden, but while doing that each staff function today also has to cope with expanding its role in its specialized area. The reason is, of course, that many of the same external events that are impacting the generalist are also taking more and more of the time of the specialists.
>
> For example, the personnel function has to cope with ERISA, EEO, and OSHA, just to name part of the alphabet, but in a somewhat broader context it must deal with changing lifestyles and values. The law function clearly has broadened, reflecting the many changes in legal relationships between business, government, and society at large. There are changes in contractual relationships, antitrust, internal trade agreements, and many more. The planning function must cope with discontinuities that make long range planning extremely difficult, if not hazardous. That's especially true abroad. Scores of corporations must carry out risk-versus-reward analyses every day regarding international business. Sperry, for instance, has operations in thirty-three countries and markets in about one hundred and forty. Only about twenty-five of them are democratic market nations. You can appreciate the challenge of just that one reality.

The finance staff today has to cope with situations that ten years ago we never thought would come to pass: double-digit interest rates. It's a problem compounded by the collapse of the equity markets. There are new and involved restrictions with repatriation and use of capital and so on.

The accounting staff has to cope with a host of new reporting standards that impinge on the flexibility of the way we do business. And, of course, the corporate communications functions have been changed, too. More and more, the communications role is becoming one of providing general intelligence about the environment for our operations. Of course, I could go on listing the ways in which corporate managment functions are changing in detail and in scope. Many are indeed social inventions. All that I've said is mainly to emphasize the point that the need for the entrepreneur has never been greater. We need new initiatives, new approaches, new vigor.[46]

But, having said this so well, Mr. Lyet still does not emphasize how much more difficult all these constraints have become for the individual innovator, and that the smaller companies simply cannot afford the overheads he describes as being essential to the functions of the large corporation in today's milieu. Nor does he emphasize that so many big corporations, for all these reasons, have become increasingly less innovative and entrepreneurial. Basically of course, it boils down to the quality of management, which varies from company to company, and in its responses to the external problems I have been describing.

The Hayes-Abernathy Critiques

A criticism of many corporate managements was articulated by two Harvard Business School professors in articles in *The New York Times* and the *Harvard Business Review.* The latter article is significantly entitled "Managing Our Way to Economic Decline."[47] It is not desirable to repeat here the extensive discussion they present on their theme, in which they basically declare that the conventional explanations for America's relative decline in innovation are inadequate and do not take into account "the failure of American managers to keep their companies technologically competitive over the long run."

They place a great deal of blame on a

new set of management attitudes and practices in the United States that have led us to downgrade our ability to compete *technologically* with other countries. American companies increasingly turn to people with financial and legal backgrounds to fill their top positions. Today, the percentage of chief executive officers of the 100 biggest corporations with these backgrounds is up 50 percent from its level of 30 years ago.

They describe the philosophies of this new breed of professional manager, and then say:

> as appealing as these ideas may sound . . . a fallacy underlies them all: that there is no need to invent, build or develop anything yourself. They all assume that, given the capital and good financial management, anything of value can be bought and any problem can be sold. . . . Building dynamic, organically growing companies requires a different breed of manager . . . who knows his markets and technologies well enough that he can break the old rules through innovation; not an executive who is preoccupied with the *process* of management.

As they say, instead, the strategic response of top management to the perceived deterioration in product and process innovation—substantially caused by financial criteria—has typically been financial: "either divest the problem business or 'buy salvation' through acquisition."

Of course they recognize that many of these trends in management are themselves the consequences of the disturbing times in which we live. They say, "We do not believe that the technological issues facing managers today can be meaningfully addressed without taking into account marketing or financial considerations; on the other hand, neither can they be resolved with the same methodologies applied to these other fields." They add, "The establishment of a strong competitive position through in-house technological superiority is by nature a long, arduous, and often unglamorous task. But it is what keeps a business vigorous and competitive."

Another very interesting and quite comparable analysis was published by Professor James B. Quinn at Dartmouth[48] and William P. Sommers[49] of Booz, Allen and Hamilton.

In my experience with many companies both foreign and domestic, I have found that these critiques are very telling.[50] The true international situation for us is very complex indeed and only one, albeit major, aspect of it is technology, but management and management strategy are indeed of critical importance.

A fascinating recent insight into this same problem was provided in an article in Mobil Oil's successful oil exploration in the October 13, 1980, issue of *Business Week*. Their chief executive officer, Rawleigh Warner, himself of financial background, is quoted as saying: "The financial people . . . did a disservice to the exploration people. . . . The poor people in exploration were adversely impacted by people in the company who knew nothing about oil and gas." Fortunately for Mobil, as the story shows, the pressures of competition forced them to recognize the faults in their strategy and to change it in time.

The Malpas Discounted Cash Flow Approach

Because of his unique experience both at Britain's Imperial Chemical Industries and as our president, Robert Malpas has had an unusual opportunity to reflect on many of these factors and critiques. His recent paper in *Chemtech*,[51] which was published also by *Chemistry and Industry*, has some profound observations and conclusions to which I refer only because I could not do justice to them in this summary. He starts with his long-felt view that DCF was the "enemy of strategy" but in the paper shows that if a proper long-term perspective is maintained it need not be so. In other words, he has turned the methods of the financially oriented managers around and showed how they can be used to incorporate a long-range risk-taking technological strategy. He also cautions against the easy tendency to debottleneck and make only marginal improvements while putting off fundamentally new investment for too long.

> Excessive debottlenecking not only increases the average age of assets but also puts back the date when new capacity can be justified. And it is new capacity using the latest technology which makes the greatest contribution to productivity, to energy saving, and to environmental improvement: a contribution which is essential to reduce inflation. Further, excessive effort on debottlenecking can divert valuable research resources away from seeking major advances.

The Landau-Malpas "Two Culture" Management Approach

It must be clear from what I have written so far that I too believe that special attention to motivation and organization are required for innovation, large and small. Of course, I know our own organization the best, although I naturally have developed some working knowledge of the many large (and a few small) chemical and oil companies that we have worked with over the years. This chapter is not the place to enlarge upon this subject. But I have written about my views elsewhere[52] as has our president Mr. Malpas.[53] However, a brief summary of some of our observations and experiences regarding organization for innovation and strategic planning might be as follows (these fall into the category of social innovations, but of a relatively conservative character).

One of the problems (of large professionally managed corporations with wide ownership diffusion) is to maintain and enlarge an entrepreneurial longer-range spirit and vision simultaneously with the employment of the systematic cost-benefit approach to decision making. That is the hallmark of good contemporary professional management. In order to overcome the obstacles I have described, this probably requires the establishment of two different cultures within the same organization, staffed by different types of people. In essence, the large organization must seek to imitate the successful looser pattern of an entrepreneurial smaller company in order to develop similar skills and ultimately successes. Conflict must be avoided between the full utilization and improvement of existing technology and the creation of the really new. The first must above all be adequately profitable and must finance the second, which is in turn needed for the company as a whole to remain profitable in the long run. It is because they have difficulty in making this separation that the venture capital subsidiaries of some big companies have not been outstandingly successful.

The boards of directors must prod the managements to provide for long-run change,[54] and assist this process by providing incentives for executives which do not depend on short-range financial accomplishments. For example, compensation for retired executives for a period of years might well depend in part on the performance of their companies during these post-retirement years. *A portion of the capital budget should be set aside for risky longer-range investments, to go along with a significant percentage of basic research of the more adventuresome kind.* The R&D department of a company exists for the very purpose of upsetting the assumptions made by its strategic planning group, since the R&D officer is devoted to technological change whose outcome cannot often be predicted. There is need for a structure closely tied to top management which integrates these divergent cultures into an entrepreneurial strategy (including technology).

But such an arrangement is difficult for many companies, both because the CEO increasingly is becoming financially oriented, and because his tenure is on the whole not very long in terms of the years needed for strategic changes. At the same time he must deal in an increasingly adversarial capacity with the primarily legally trained government officials and politicians, as Mr. Lyet points out. Thus, there is now a three-way tension between lawyer-financier-technologist in each of these centers of intellectual and economic activity which somehow must be bridged by the CEO—whatever his training—together with the board; despite all the short-term pressures, the company must increasingly have a long-range strategy. A look at the companies currently in the Dow-Jones averages versus those 40 years ago will show little repetition. Thus, the fate of the company itself is ultimately at stake in such strategic planning. If this is indeed true, it must accept the penalties of failure and enjoy the rewards of success, both without undue involvement by government.

An article in *The New York Times* about the problems of one such real

CEO is most pertinent to the theme of this chapter[55] and should be read also with the comments on the technology aspects of his company cited in Young.[56] These problems of publicly held corporations are also complicated by the fact that the increasing proportion of outside directors is coming primarily from people of legal or financial backgrounds, rather than from technological activities. Thus boards have audit committees, pension committees, etc., but very few have a technological audit committee. The company just referred to above happens to be one of them. And although a recent article[57] urges a strategic planning committee of the board to work closely with the CEO, this is not practiced very extensively for similar reasons. Such a situation exists in many companies, even with the high technology industries referred to earlier in this chapter.

There is need for greater flexibility and better long-range planning in technological strategy. Our foreign competition does not stop. The Japanese efforts in electronics and semi-conductors are already well known; more recently, they have been applying their techniques to chemicals. As it happens, it is also an area in which Halcon is now very active, namely, the chemistry of carbon monoxide. In fact, we announced this year our newest breakthrough and its first commercialization: the carbonylation of methanol to produce acetic anhydride, vinyl acetate, acetaldehyde, etc. The work stemmed from the realization that ethylene prices were heading upwards very rapidly, and that carbon monoxide from various lower-cost carbon-containing feeds such as coal, residual oil, gas, biomass, etc., could be made an important source for synthesizing higher molecules. Tennessee Eastman has just announced a large acetic anhydride plant from coal which will contain Halcon technology in part, and Halcon retains exclusive rights for the valuable knowhow that will result. Thus, we are moving quickly to gain a commercial edge and to start on our learning curve. Plans are advancing rapidly to apply this technology and experience to vinyl acetate manufacture in at least three countries.

But the Japanese are pursuing another approach. For the first time in their chemical industry, the government is funding an extensive research program in carbon monoxide chemistry, mostly conducted by a group of private companies who will make their results available to all. What will emerge from this is problematical, as it visualizes no plants for a number of years. But we in the United States have no time to lose!

A recent lengthy insert in *Scientific American*[58] gives an excellent insight into some of the Japanese strengths and weaknesses in innovation, which have in fact been more social than technological. It would be an error to draw too many lessons from Japan for ourselves because of the very substantial differences between our societies. Thus, for example:

1. Japan has practiced a discriminatory protectionism against foreign capital and goods for years, with our tolerance.
2. There are few union problems in Japan.
3. By practicing early retirement at 55, Japanese workers are forced to make jobs for younger people, thus redefining unemployment.
4. There is little social security, so that private savings must be high and this has permitted long-range investment in industry at low return rates in the short run.
5. A single party (conservative) has dominated the scene since the war, with an ineffective opposition, so that policies can be developed on a longer term basis.
6. The Japanese population is basically homogeneous ethnically and religiously, with resultant fewer social tensions.
7. Population control is practiced by widespread abortion, among other means, and immigration is rigidly restricted.
8. Pressures for conformity to the group are overwhelming, and individual initiative is difficult to develop in a "consensus" system.
9. There is no effective antitrust experience or tradition, and large businesses have all kinds of arrangements with each other and their government.
10. The Japanese constitution was recently written and imposed by General MacArthur, thereby freeing the Japanese from many traditions and customs of the past which our society cannot readily do.
11. There are only about 10,000 lawyers in Japan, vs. our 500,000, hence an absence of regulations and litigiousness.

CONCLUSIONS

I have devoted much space to getting into the entrepreneur's world—past and present, individual and corporate—in order to demonstrate the reality of just a few very basic lessons I have learned over these past 34 years:

1. The entrepreneur, particularly the technological innovator, functions best in an atmosphere of freedom, in a decentralized system.[59] It is probably also true that this grass-roots innovation is the best method for most social innovation too. Certainly, large government-sponsored programs of specific national importance have been successfully implemented through the private sector firms in the past (World War II synthetic rubber, the atomic bomb, the space program), but for these to

succeed in the world market under the pressure of competitive forces, the action of the "invisible hand" is infinitely more effective. It is very doubtful whether the currently fashionable concept of technology assessment has any real long-range value, as studies of the past (such as by Professor Rosenberg, or by Professor Hughes) have demonstrated. On the other hand, it is absolutely necessary, in view of the speed with which science and technology change, to develop greater technical literacy among the general public and to communicate honestly with them about the scientific work that is being done, while at the same time improving our engineering education generally.

2. The entrepreneur, whether individual or corporate, requires an external climate marked to the maximum degree by steadiness of course in the society around him, although he usually is prepared to cope with a reasonable degree of surprises. However, the perils of the entrepreneur's career itself in a competitive world are amply able to provide him with surprises, so that he doesn't need any additional burdens from governments or segments of society acting through government.

3. Technological innovation has been *the* major driving force for the growth of the American economy for a very long time. As Charles Schultze has put it:

> The final virtue of market-like arrangements that I wish to stress is their potential ability to direct innovation into socially desirable directions. While the formal economic theory of the market emphasizes its ability to get the most out of existing resources and technology, what is more important is its apparent capacity to stimulate and take advantage of advancing technology. Living standards in modern Western countries are, by orders of magnitude, superior to those of the early seventeenth century. Had the triumph of the market meant only a more efficient use of the technologies and resources then available, the gains in living standards would have been minuscule by comparison. What made the difference was the stimulation and harnessing of new technologies and resources.[60]

What Schultze does not say is that this market triumph, occurring only in Western countries, did not therefore arise in the ancient civilizations of the East, where the idea of progress did not exist, and which to this day have living standards far below those in the West, so that the social milieu is after all the decisive factor.[61] This feedback ability of the American economy, if not stifled, is its greatest strength, because it is self-correcting. Thus, the real priority belongs to technological innovation. It is government's job to set the course and then get out of the way! If there are no foreign "surprises," the developing national consensus as described above might permit a gradual improvement in our national condition during the 1980s.

4. The driving force behind technological innovation is money—the lure of making it. For the individual, that motivation is extremely clear, even though, of course, there are other motivations as well (desire for independence, hope of peer group esteem, etc.). For the corporation to innovate, given the problems I have described, there is no better recipe than a profitable business. A good profit record engenders a willingness to take risks, and risk-taking lies behind all innovation. Similarly, Professor Mansfield[62] points out that the higher social than private rate of return for innovation cannot efficiently be addressed by direct government intervention, but rather by improving the profit-making climate for innovation.

Thus, in my opinion, there has to be a recognition at the highest levels in the United States that a trade-off is necessary between encouraging new risk-taking wealth among corporations and individuals—wealth that will translate into investment—and the desire for equity and redistribution of income. Considering the great need for breakthroughs, as well as improvements, this trade-off will have to be settled largely in the direction of wealth creation and new entrepreneurial incentives by tax reduction and regulatory reasonableness (including elimination of the concept of zero-risk). *Nothing else will realistically work.* This wealth formation is not being encouraged soley for its own sake, but because it is the only way the country's economic and social welfare can be improved for all the people in a free society, as Charles Schultze has pointed out. In essence, we must by these means concern ourselves with enlarging the national pie, not in how equally we slice up a static or shrinking pie, a process fraught with social strife. We have been taking more out of the economy than we have been putting in for the last twenty years, so that our priorities must be changed accordingly. Increasing our productivity will allow us to increase the size of the pie as well as the size of the slices, but it will take patience and persistence to change our habits of the recent past. Most especially, we cannot continue to rely solely on credit restraint and monetary policy with cruelly high interest rates to reduce inflation, when government continues to spend so much more than it takes in, and wages are rising so much faster than productivity. These are the real causes of our persistent inflation.

It is absolutely incredible how, in today's perverted environment, making a lot of money seems all right, indeed admirable, if the individual involved is a sports celebrity, an actor, a rock star, or a disco singer, but is bad if the person is the chairman of General Electric or any other major company that employs hundreds of thousands of people and contributes significantly to our Gross National Product. The same opprobrium has attached to the entrepreneur who seeks capital gains for his efforts, say, to harness the biological technologies which may be of incalculable value to humanity. Why is this so, as it seems to contradict the prevailing egalitarian trend of which I have written? In my opinion, because the egalitarians have persuaded the voters that they are just as good as all these executives and entrepreneurs but the voters are willing to admit that they can't be like Jane Fonda or Reggie Jackson. Of course, this is romantic nonsense, but it underlines how distorted our risk-reward system has become, and how it must be recharged all up and down the ladder. There must once again be rewards for success, not failure, for merit and performance. Rewards not because "Joe next door gets more so I want the same," rewards for the hardworkers, not the indolent. This is particularly important because some of the nineteenth century motivations, such as religion, the family, the work ethic, and patriotism, no longer have the force they once had. Our experience with our voluntary defense forces is showing this.

5. Of course, in order to restore our innovative and economic growth, we shall also have to provide social inventions. Some of these are already in existence, such as "safety nets," redundancy retraining, aid in moving workers from high unemployment areas, aid in matching jobs with workers, special attention to the problem of long-term sectoral joblessness[63], and many others for which I have only sympathy, although they may well need improvement and refinement, but we must never lose sight of the above-stated basic requirements. Others will surely have to be invented and, yes, disinvented (such as price supports, monopolistic unions power, etc.), just as we were inventing and pioneering innovative financing systems, antitrust approaches to business, land-grant colleges for practical education, and many other ideas of the past which helped shape the innovative milieu of which I have written in this chapter. But let us not fall victim to the fallacy that we can shore up the declining present by certain types of social inventions (such as "trigger prices" or failing corporate bailouts, where inefficient managements and acquiescence to excessive labor demands, usually combined with falling productivity improvement, should be penalized by paying the price of their transgressions, not rewarded by the general taxpayer); in our efforts to do so we will only further succeed in weakening the real requirements of our society for innovation, competitiveness, and growth. It would be a disaster for us in the long run if we opted for a cozy three-way "arrangement" between big government, big labor, and big business. That path will lead to dictatorship.

6. We know that the technological innovation of the past was not without its costs, as a reading of Dickens, Dostoevsky, or Balzac makes blazingly clear. They also illuminate, however, how terrible was the general poverty only a century ago, and how far our progress has been

since then in spite of the negatives. In fact, I can vividly recall how far we have come since the 1930s when I entered college. A visit to Calcutta, or Kingston, Jamaica, or the Favelas of Rio de Janeiro, or any of countless rural villages in the third and fourth worlds will tell us how far humanity as a whole still.must go, and how fortunate we have been to have been accorded our particular circumstances, for which we owe a return obligation, not only to our own less fortunate citizens, but to many others. But there is no guarantee of continuing success, as the decline and fall of many a civilization amply attests. It is up to us to profit from the lessons of the past, not superficially, but in the most profound ways. I am convinced that our high-tax, high-inflation, low-growth, low-profit economy, if long continued will, like heroin addiction, lead to the death of the patient; withdrawal symptoms will be painful and perhaps prolonged, but they are necessary for his salvation.

Notes

1. E. Mansfield et al., *The Production and Application of New Industrial Technology* (New York: W. W. Norton & Co, 1977), p. 12.
2. Applied Science and Technological Progress, a report to the Committee on Science and Astronautics, U.S. House of Representatives, by the National Academy of Sciences, June 1967, Introduction, p. 1.
3. Applied Research Definitions, Concepts, Themes, p. 21.
4. N. Rosenberg, "Technological Interdependence in the American Economy," *Technology and Culture* **20**, no. 1 (Jan. 1979): 25.
5. E. Mansfield et al., *The Production and Application* p. 72.
6. W. M. Blumenthal, Address before the Annual Conference of the Financial Analysis Federation, Bal Harbour, Fla., May 8, 1978. This address gives many statistics and useful confirmations.
7. R. H. Hayes and W. J. Abernathy, "Managing Our Way to Economic Decline," *Harvard Business Review*, July–Aug. 1980, p. 67; *N.Y. Times*, Aug. 20, 1980, p. D–2; Aug. 22, 1980, p. D–2.
9. *Economic Report of the President*, Jan. 1977, pp. 45–48.
10. R. Atkinson, Basic Research, The MacNeill/Lehrer Report televised New York, Nov. 30, 1977.
11. W. D. Nordhaus, at AAAS Second Annual Colloquium on R&D in the Federal Budget, June 15, 1977. See also E. Mansfield, Congress of the International Economic Association, Tokyo, 1977.
12. M. Friedman, *Capitalism and Freedom* (Chicago: University of Chicago Press, 1962).
13. R. Landau, U.S. Patent 2,752,362. Process for the Oxidation of Ethylene.
14. R. Landau and R. Lidov, "Ethylene Oxide," *Ethylene and Its Industrial Derivatives*, ed. by S. A. Miller (London: Ernest Benn Ltd., 1969), Chapter 7, pp. 513, et seq: R. Landau, D. Brown, A. Saffer, and J. V. Porcelli, Jr., "Ethylene Oxide Economics; The Impact of New Technologies," presented at the 60th annual meeting of AIChE, New York, Dec. 1967, printed in *Chemical Engineering Progress*, March 1968, p. 27.

15. E. G. G. Werner, (Managing Director of Royal Dutch/Shell group of companies), "A Change of Gear for Petrochemicals," address to the Technological Club, Delft, 15 September 1978; *Chemtech,* July 1980, p. 430.
16. R. Landau, "The Chemical Process Industries in International Investment and Trade," *Proceedings of the Technical Session at the Eleventh Annual Meeting April 23-24, 1975,* National Academy of Engineering.
17. T. J. Teece, "Technology Transfer by Multinational Firms: The Resource Cost of Transferring Technological Know-How," *The Economic Journal* **87** (June 1977): 242-261.
18. R. Landau and A. Saffer, "Oxidation of Aromatic Hydrocarbons," presented at the Modern Chemistry Symposium sponsored by *Modern Chemistry in Industry,* Eastbourne, England, March 11-14, 1968; R. Landau and A. Saffer, "Development of the M-C Process," *Chemical Engineering Progress,* **64,** no. 10 (October 1968): 20-26.
19. R. Landau et al., U.S. Patent 2,855,430, "Process for the Preparation of Diisopropylbenzene."
20. A. Saffer, U.S. Patent 2,833,817, "Process for the Preparation of Terephthalic Acid"; R. Landau and A. Saffer, U.S. Patent 2,833,818, "Process for the Preparation of Terephthalic Acid"; U.S. Patent 2,858,334, "Preparation of Phthalic Acids."
21. A. Saffer and R. Barker, U.S. Patent 2,833,816, "Process for the Preparation of Aromatic Polycarboxylic Acids."
22. W. J. Casey, *Brooklyn Law Review,* Spring 1977.
23. R. Landau, "Chemical Industry Research and Innovation," in *Innovation and U.S. Research,* Symposium Series 129, American Chemical Society, Washington, D.C., 1980; R. Landau, "Halcon International, Inc., An Entrepreneurial Company," Newcomen Society dinner, June 21, 1978, Publication No. 1088, Princeton University Press.
24. T. O'Hanlon, "Less Means More at Firestone," *Fortune,* Oct. 20, 1980, p. 114.
25. David E. Kostoff, *The Mellons,* (New York: T. Y. Crowell, 1978).
26. *Newsday,* June 18, 1978, review by Harry Steinberg.
27. G. Bylinsky, *Fortune,* Dec. 1977, p. 76.
28. R. Landau, Hearings of the Joint Economic Committee, Oct. 1, 2, 3, 10, 1974, p. 112; Hearings before the Committee on Ways and Means, House of Representatives, Jan. 30-Apr. 24, 1978, p. 3026; "Entrepreneurship in the Chemical Industry and in the U.S." in "Innovators and Entrepreneurs—An Endangered Species?" National Academy of Engineering, Feb. 1978; also as participant in study by the National Research Council, "The Impact of Tax and Financial Regulatory Policies on Industrial Innovation," Washington, D.C., 1980, prepared by J. J. Cordes.
29. E. Mansfield, "Research and Development, Productivity, and Inflation," *Science* **29.** (5 Sept. 1980): 1091.
30. G. F. Will, *Washington Post,* Dec. 11, 1977.
31. "Investment Strategy Highlights: Overview: Are Stocks a Bargain?" A discussion of the Equity Risk Premium, Nov. 1977, Goldman, Sachs & Co.
32. Alan Greenspan, *The Economist,* Aug. 6, 1977.
33. R. Landau and A. I. Mendolia, "International Chemical Investment Patterns Reviews," *Chemistry & Industry,* Nov. 19, 1977, pp. 902-910.
34. "Capital Gains Taxes," *The Economist,* Oct. 4, 1980, p. 85.
35. *The Business Roundtable,* 1977.
36. C. L. Schultze, "Regulation," *AEI Journal on Government and Society,* Sept./Oct. 1977, pp. 10-14.
37. R. Salomon, "Second Thoughts on Going Public," *Harvard Business Review,* Sept./Oct., 1977, p. 126.
38. "Industrial Innovation and Public Policy Options—Report of a Colloquium," National Academy of Engineering, Washington, D.C., Oct. 1980.

39. *Science,* May 6, 1977, p. 636.
40. R. Amann, J. M. Cooper, R. W. Davies, eds., *The Technological Level of Soviet Industry* (New Haven: Yale University Press, 1977), Chap. 6: The Chemical Industry.
41. E. A. Gee and C. Tyler, *Managing Innovation* (New York: John Wiley & Sons, 1976).
42. R. Malpas, "Chemical Technology—Scaling Greater Heights in the Next Ten Years?" *Chemistry and Industry,* 5 February 1977, p. 111.
43. R. Landau, D. Brown, J. L. Russell, J. Killar, "Epoxidation of Olefins," presented at Symposium on *New Concepts and Techniques in Oxidation of Hydro-carbons,* 7th World Petroleum Congress, Mexico City, April 1967.
44. R. Landau, D. Brown, G. A. Sullivan, "Propylene Oxide by Co-product Processes," *Chemtech,* October 1979, pp. 603–607.
45. R. Landau, "Halcon International, Inc.—An Entrepreneurial Chemical Company," The Newcomen Society in North America, Princeton University Press, Princeton, N.J., October 1978, publication no. 1088.
46. J. P. Lyet, "The Role of the Entrepreneur in the 1980s," Wharton Entrepreneurial Center, The Wharton School, University of Pennsylvania, March 1980.
47. R. H. Hayes and W. J. Abernathy, "Managing Our Way."
48. J. B. Quinn, "Technical Innovation, Entrepreneurship, and Strategy," *Sloan Management Review,* Spring 1979, p. 19.
49. W. P. Sommers, "Managing Technology: Top Management Must Regain the Initiative," *The International Essays for Business Decision Makers* **IV,** (1979): 10.
50. R. Ball, "Europe Outgrows Management American Style," *Fortune,* Oct. 20, 1980, p. 151.
51. R. Malpas, "Meditations on Maturity," *Chemistry and Industry,* Dec., 1979, *Chemtech,* Sept., 1980, p. 558.
52. R. Landau, "Halcon International, Inc., An Entrepreneurial Chemical Company," address delivered at the 1978 New York dinner of the Newcomen Society in North America, June 21, 1978. Newcomen Publication no. 1088, Princeton University Press.
53. Malpas, "Meditations on Maturity."
54. S. M. Felton, Jr., "From the Boardroom," *Harvard Business Review,* July–Aug. 1979, p. 20.
55. I. Barmash, "America's Most Influential Jones," *New York Times Magazine,* Sept. 16, 1979, p. 34.
56. L. H. Young, "To Revive Research and Development," *Business Week,* Sept. 17, 1979, p. 27.
57. W. W. Wommack, "The Board's Most Important Function," *Harvard Business Review,* Sept.–Oct., 1979, p. 48.
58. J. C. Abegglen and A. Etori, "Japanese Technology Today," *Scientific American,* Sept. 1980, Special Section Advertisement, pp. 15–30.
59. E. Mansfield, in "Innovation and U.S. Research," ACS Symposium Series 129, Washington, D.C., 1980, "The Economics of Innovation," p. 95.
60. Schultze, "Regulation."
61. R. Nisbet, *History of the Idea of Progress* (New York: Basic Books, 1980); H. E. Meyer, *The War Against Progress,* (New York: Storm King Publishers, 1979).
62. Mansfield, "Research and Development."
63. B. C. Clark and L. H. Sommers, "Unemployment Reconsidered," *Harvard Business Review,* Nov.–Dec., 1980, p. 171.

chapter five

THE HUMAN SIDE OF TECHNOLOGICAL INNOVATION: LABOR'S VIEW

Iris J. Lav and Stanley H. Ruttenberg

American labor recognizes the value that technological innovation can confer on the economy. Labor leaders know and understand the links between innovation and productivity, inflation, and economic growth. Labor knows well the importance of innovation because workers have been hurt and frustrated by lack of innovation in their companies, especially when it is linked to elimination of jobs because an industry is no longer competitive. But labor also has legitimate concerns about the manner in which innovation is implemented and the direction it takes. While innovation may lead to a better standard of living for most workers, those workers who are directly affected by innovation have needs that can be met only by the humane social rules and institutions with which we surround and support the process of innovation. How we go about the business of innovation and how well we meet the human needs of labor will become a critical consideration in whether the society can proceed with and reap the fruits of new technology.

Labor can oppose innovation, become a willing partner of business in the process of innovation, or make a variety of responses between these extremes. The key question is, of course, "what makes the difference?" What structures, essentially social ones, must surround innovation in order for labor to view it positively? The elements of those structures in their barest form are:

1. Good communication and cooperation between labor and management
2. An equitable sharing of the fruits of innovation and increased productivity with the workers implementing the innovation
3. Adequate compensation of displaced workers and programs to assist their adjustment and reemployment
4. Programs to assist affected communities and victims of secondary unemployment as a result of innovation and change
5. A full employment economy with a variety of opportunities for displaced workers

These conditions for the smooth acceptance of change caused by innovation have been known for at least forty years; there is nothing new or innovative about them. They were articulated by John L. Lewis in the early days of the introduction of machinery into the mines. But what has been sorely missing in the intervening years is widespread acceptance of the importance of these conditions. Perhaps today, with the fate of the U.S. economy apparently more and more dependent upon the rate of innovation and technological change, new and creative social inventions can be found to meet these conditions.

Labor's Response to Innovation

There are a wide range of possible responses of labor to the introduction of an innovation. At one end of the continuum lies the conception of labor as attempting to obstruct innovation. A few well-publicized cases such as on the railroads or in newspaper printing where workers have fought against new technology—or more properly against inadequate programs to compensate for its effects on their jobs—have reinforced this image.

At the other end of the continuum lies those situations in which labor has accepted the need for change and has created or taken part in innovative structures to make optimum use of new technology. The seagoing maritime industry is an example of this, where the rapid and dramatic automation of ships and increase in ship sizes in recent years has been accompanied by cooperation from the seagoing unions and the establishment of union-supported schools for the training, retraining, and upgrading of skills necessary to operate the newer ships. The operating engineers and the printing trades unions are similarly involved in training.

Another example can be found in the construction industry where, historically, technology and change have been slow to take hold. Faced with a decline in union construction in St. Louis, the unions got together with contractors, architects, engineers, suppliers, and builders to form the cooperative program known as *PRIDE*.[1] That program includes many types of agreements and pledges by all parties, including pledges on the union side to use the new technologies of building to increase productivity and hold down costs. These include the utilization of modular housing and prefabricated materials, mechanized pipefitting, and spray painting. As a result of this cooperation, productivity increased and costs of construction went down. Five other cities have similar programs. The advantages of these programs lie both in their initial contributions to the adaptations to change and in their establishment of structures through which future changes can easily be implemented.

It is even possible to look one step further at the positive end of the continuum and find a few cases in which technological innovation as well as supportive social invention has stemmed from shop floor creativity, where intimate knowledge of the production process has suggested both the need and the means for such invention and innovation. A few recent examples of small, but important inventions and innovations of this origin appeared in a recent *Wall Street Journal* article about "quality circles," in which worker suggestions about new types of tools or better ways to design parts resulted in monetary savings to the companies.[2]

In between these extremes of response may be found the type of acceptance of the technological change which comes in exchange for a "buy out" or a quid pro quo, usually agreed upon through collective bargaining. This type of response may be more or less grudging or accepting, may be readily obtained or obtained only through protracted negotiations, and may or may not leave elements or residual hostility on either side. Examples of this type range from the longshoremen's fifty-mile container rule to the much more common agreements that any workforce reductions will occur through attrition, early retirements, and similar mechanisms. Finally, there is the response that is called "willing acceptance" in the literature. In this response, the change is readily agreed to or not opposed, because of immediate satisfaction with the adjustment procedures, because the change is perceived as inevitable, or a variety of similar reasons.

The cases of "willing acceptance" are those that never receive any publicity, but the empirical evidence suggests that they have been by far the most common responses of labor unions to technological innovation—found in nearly half of the more recent cases studied. But the other responses exist, both positive and negative. Evidence shows that the response to innovation and technological change has been opposition, at least initially, in nearly one-fourth of the cases.[3] While this opposition was not always successful and rarely permanent, it is bothersome and expensive both to the innovator who wants to make the change and also to labor, who clearly must feel pushed into a corner to make such an extreme response.

With possible responses ranging from opposition, to acceptance in exchange for a "buy out," to willing acceptance, to creative acceptance, or enthusiastic participation in the innovation process, individual companies, governments, and societies must find the means to ensure responses on the positive end of that continuum. As we will see in the following section, unions are committed to the concept of innovation and technological change. That commitment can be a benefit to society, business, and labor, if only enough creativity, good-will, understanding, and resources exist to create the conditions under which labor can afford to respond willingly and indeed enthusiastically.

Toward a Positive Response

There are four levels on which the types of social structures that exist can be important. They are (1) within the union itself, (2) within the company, (3) within the economy and the government, and (4) within the society—particularly within what is perceived as the corporate society.

The Union

Unions are moving ahead to create the climate for acceptance of innovation and technological change within their leadership and membership. It is becoming a topic of increasing importance in union thinking and actions.

A recent report and recommendations on pension policy developed by our firm and adopted by the AFL-CIO Executive Council in August 1980, exemplifies the type of commitment that is now in the air. The first of those recommendations advocated the use of funds from pension plans, funneled through a Congressionally created public institution, "to provide the capital necessary to stimulate the development of new job-creating industries and/or to support modernization of existing industries; both with a view toward maintaining the ability to compete in U.S. and world markets and, at the same time, to increase productivity."[4] Not only are the unions specifically recognizing the need for innovation and increased productivity in U.S. industry, they are willing to use some portion of the deferred compensation of union workers to achieve that end.

A variety of individual unions are also looking at the need for innovation and the role that labor can play. The Communications Workers of America is studying its industry in order to predict the course of future technological change. It wants to be able to prepare its members in advance to accept technological innovations, as well as participate in training and retraining their members to be ready for the new jobs that innovation will bring. The Graphics Arts Union has an ongoing training and retraining program in sixty cities, where apprentices receive state-of-the-art training, and where "journeymen are upgraded or are being retrained for jobs in other areas of the industry when their former jobs have become outmoded."[5]

In 1979, the Department for Professional Employees of the AFL-CIO sponsored a conference on "The Impact of New Technologies on the Work Force."[6] And the Marine Engineers' Beneficial Association, the union that represents licensed engineers aboard U.S. ships, is conducting some of that reverse transfer of technology about which so much is said these days. They are using their pension funds to form a corporation to build and operate a certain type of highly advanced ship that is used in Europe but not in the United States.[7]

These are just a few diverse examples of the fact that innovation and change are very much on the minds of unions today. Unquestionably, unions are now looking for ways to create and to participate in change.

There is some suggestion in the literature that certain unions, by virtue of their organization and structure, are better able to respond positively to technological innovation than others. It has been said that stronger unions, with more control vested in the national or international office, are better able to deal with the "big picture" of the benefits of such innovation. It is said that these unions are better able to initially bargain for satisfactory adjustment mechanisms that prevent an impasse later in the process and thus better able to present a change to their members in a manner in which it will be accepted.[8] This may or may not have been true historically. It certainly is true that unions can rarely make an adequate response when they are dealing from a position of weakness.

Whatever the historical pattern has been, today most unions are fully aware of the need for technological innovation to revitalize the economy. Most unions are ready, able, and willing to creatively participate in that revitalization—*if* they had reason to expect responsibility in the face of such innovation on the part of companies, *if* they could be sure of adequate programs to assist those workers who are displaced, and *if* they didn't have to spend a good deal of their time and resources fighting against a rising tide of antiunionism in corporate society.

The Company

There are two types of corporate responsibility related to technological innovation and technological change. First, the company has the obligation to share the fruits of such innovation with the workers. If the innovation leads to increased productivity, some of that greater productivity should be realized in the form of higher profits for the company, but some should also be shared with the workers in the form of higher wages. Ideally, the greater productivity should be shared three ways—among profits, wages, and lower prices for consumers—for it is in that way that innovation leads to a better standard of living for society.

Second, if the innovation does result in the displacement of some workers, the company must bear some of the responsibility for the adequate compensation of these workers. As will be discussed later, the society, through the government, also should hold some of the responsibility for the training and reemployment of those displaced workers— it is not entirely a company responsibility. Just where the responsibility divides is an important matter under discussion in Congress and elsewhere today. But at the least, the company benefitting from the innovation must bear the responsibility and the cost of adequate severance pay based on length of service, the age of the worker, and the period of unemployment. The company should also be responsible for maintenance of worker benefits such as health and life insurance and pension contributions and even assistance with mortgage payments during a reasonable period from the time of severance during which the worker may be unemployed. And depending on the circumstances, some changes in pension rules may also be desirable in terms of early vesting and early retirement provisions.

The remainder of the needs of workers adversely affected by innovation should be the responsibility of society through the government. These needs include training and retraining, relocation assistance, assistance to communities, and assistance to workers who are displaced through the secondary effects of innovation. More will be said about that later.

There are two ways in which the limits of corporate responsibility can be defined. One is through collective bargaining. The other is through legislation which would at least establish a floor on the responsibility of companies. Both may be necessary. But from whichever source the responsibility derives, the elements of communication and trust are necessary to make the implementation of the innovation a smooth transition.

The quality of the relationship between management and workers—the ability to communicate and the availability of established channels, flexibility and mutual trust—are all critically important. That has been said so often it sounds trite, but for all of the articles written extolling such virtues, the virtual means of such essential communication and trust are still in their infancy and are much in need of creative, practical, application.

In a situation of technological innovation, effective communications are needed to provide early notice of the innovation and its potential effects on the workers, to discuss the need for the nature of the innovation, to foster understanding of the human impact of the innovation, to discuss measures that will be taken to offset the adverse effects of innovation on particular workers, and to foster understanding of the innovation process. The bargaining table is the proper focus for discussion of most of these matters, at least initially. But further and wider discussion in labor-management committees can both demonstrate the concern of management for the human problems innovation often entails, and allow workers a meaningful participation in the process. Through labor-management committees, workers can make their own assessments of the meaning of the new technology to them, give their insights on possible problems that might arise on implementation, and make alternative suggestions. In ongoing committees of this type, the knowledge and ideas of workers about additional places where new technology and/or social inventions are necessary to improve productivity can also be tapped to flow upward.

One recent example of a labor-management committee in a service industry illustrates the type of upward flow of social inventiveness that is possible. This example involves the A & P stores in southeastern Pennsylvania and the Food and Commercial Workers. Briefly, the A & P was ready to close sixty stores in that area because of poor productivity and profitability. Under the threat of crisis, labor-management committees were formed. Together they worked out solutions to problems as diverse as employee attitudes, store budgeting, scheduling, advertising, and responses to competition from fast food chains. Profits have improved and the entire division is projected to be in the black next year. The turnaround has occurred despite depressed economic conditions in the area. The grievance level has fallen from a very high one to virtually nil. And most importantly, the company has agreed to make needed investments in modernization of several of the stores in the area.[9]

This is a small local success story of cooperation to bring about needed changes—in this case changes of technique rather than technology of operation. But it should be noted that the retail food industry as a whole is one that has many problems ranging from the price of energy to the changing life styles of families. It also is in the throes of the technological innovation of electronic scanners. And while many problems remain unsolved, there is an industry-wide Food Industry Joint Labor-Management Committee providing an important climate and forum for their solution. There are other examples of local and national committees, but they are not numerous and very few are focused on the issue of innovation in technology.

There is often reluctance from both management and labor to establish such committees. To simplify somewhat, management has shown reluctance to share its information—in this context, advance information on future changes—and what they see as their perrogatives with labor. In some cases, management is reluctant to acknowledge that labor has important contributions to make. For its part, labor is concerned lest the committees become a substitute or an avenue for bypassing collective bargaining and the union organization itself, which they cannot be. As a result, such committees have only been established in a handful of the situations in which they could be helpful. It should not take a crisis, as it often does today, to bring on this type of essential cooperation. This is an area, in our opinion, where creative social inventions that lead to improved communications are badly needed.

Labor often has a very different perception of the reasons for a particular change or technological innovation than those stated by management. We have said that in the vast majority of cases unions are not opposed to increasing productivity. They are not opposed to technological innovation because they can see the need for both a competitive industry and a healthier economy. But they do not always believe that proposed changes foster those ends. Particularly when whole plants are closed or when substantial amounts of the work of a plant is transferred to another plant, either in this country or abroad, labor often feels that a profitable plant has been closed down by an impersonal corporation to generate a tax benefit, or that a move to the Sunbelt is being made to avoid bargaining with the union—without ever having given the union a chance to resolve whatever mutual problems might exist at the current plant.[10] Even when there is only a technological change taking place within a plant, workers may perceive the reason for that change as an attempt to define a group of jobs to be outside of a bargaining unit.[11]

If these are not the reasons change is occurring, then improved communcations become all important. Creative use of early notice with full explanations, participatory committees, and good faith bargaining should be able to establish the need for the change and delineate the benefits as well as the costs to each party before antagonistic ideas can develop from rumor or partial information. If the reasons for the change are, in fact, those that labor considers questionable or illegitimate, then that purpose will not be lost on the workers and it isn't surprising if the response turns out to be negative.

Finally, it should go without saying, but unfortunately cannot, that the way in which employers view the work-force is critical. If they are viewed as just another factor of production that—to exaggerate only slightly—can be scrapped at will, cooperation will not be possible. Employers have to view worker adjustment costs as part of the cost of industrial innovation and not grudgingly so; a form of human resource accounting.

It might be fair to say here that scrimping on the allowance for the human costs of adjustment stems not only from individual employer predilections, but also from the way in which a company's performance is judged—almost solely on the bottom line, cents per share, profit basis. When we talk about creative social inventions to support technological innovation, we may need to think about creative new ways to judge and evaluate corporate performance. As Tom Donahue of the AFL-CIO pointed out recently, perhaps the success or nonsuccess of a corporation should also be judged on their responsibility toward their workers and the communities in which they operate.[12] We could devise specific ways of recognizing those companies that have fulfilled their responsibilities well. And the goals of profitability and responsibility do not have to be conflicting. Opposition and delay in innovation is costly to the corporation, labor, and the society. Behavior that willingly recognizes the human costs of innovation can mean better profitability in the longer view from the ability to innovate as needed.

The type of corporate responsibility and care for the human costs of innovation we are talking about is not an abstract dream; it can occur. Recent testimony by a representative of Brown & Williamson Tobacco Corporation before a Senate committee impressed many people with the manner in which they handled needed changes.[13] To summarize briefly, the company was faced with a situation in which production had to be transferred to a new, efficiently designed plant in Macon, Georgia. The old plants were located in Louisville, Kentucky, and Petersburg, Virginia. The Louisville plant was to be closed completely and the Petersburg plant substantially reduced in force. Ten different unions represented the workers at these plants. The collective bargaining agreements in effect specified that workers were to be given advance notice of any changes, a provision that had not been resisted by the company. As soon as the company realized that the transfer would be necessary, they notified the unions and began negotiations over the effects of the change. After agreement was reached, joint committees of union and management transmitted them to the employees.

The agreement contained the following elements:

1. Severance pay from 26 to 56 weeks depending on length of service
2. Change of pension plan to "rule of 70" allowing early retirement at age 55
3. Continuation of life and medical insurance for up to six months from severance at no cost to employee
4. Designation of a specific number of jobs at the new plant for Louisville and Petersburg employees, with transfer rights based on seniority
5. Comprehensive financial assistance for relocation, including pre-move visits
6. Counseling for severed employees
7. Mailings to other industries in the state to alert them to the availability of good employees
8. Assistance in preparing resumes and paid time off for interviewing
9. Company paid training and retraining for new jobs
10. A joint effort by the company and unions to secure changes in State Unemployment Compensation laws to allow collection of severance pay and unemployment benefits.

By limiting these benefits to employees who agree to leave or transfer at times specified by the company, the company is able to have an orderly transfer of operations over a three-year period. To quote the company:

> We asked for and received the cooperation of union leadership in avoiding a potentially explosive situation. What could have resulted in sabotage, wildcat strikes or slowdowns became an orderly procedure, with loyal, however saddened, employees continuing to work regularly and diligently. . . . The company, its employees and its unions were benefited by union/management relationships based on respect, trust and recognition of need.

And to quote the union side, "Brown & Williamson has demonstrated the type of company they are in their concern for their employees by agreeing to the best plant closure agreement I've ever seen. Probably the best in the country."

The Economy and the Government

It clearly is easier for any adjustment process to take place in a growing, full employment economy, when workers' concerns about the effects of industrial innovation are likely to be less. The very history of concern over the effects of such innovation mirrors the conditions in the economy. As we came out of the decade of the 1950s with high unemployment and major recessions one after another, there was a flurry of concern in the early 1960s about the problem known then as "automation." For example, the purpose of the first version of the Manpower Development and Training Act, passed in 1962, was to retrain those workers displaced by automation. As another indication of concern, a National Commission on Technology, Automation and Economic Progress was appointed by the President in 1964. But by the time that commission made its report in 1966, concern over the problem had waned. The economy was at or close to full employment, and the focus of the MDTA as well as the attention of policy makers was fixed on the special problem of structural unemployment. The problem of compensating for the effects of technology on workers was not much heard of again until the latter part of the 1970s, when a lagging economy and high unemployment has again created concern with the effects of innovation.

While displacement due to automation or innovation takes place whether there is full employment in the economy or not, the measures to compensate for that displacement can be more modest in a growing full employment economy because workers can be rapidly reabsorbed in the growth of their own or another company or industry. But since we don't have the best of all possible worlds all the time, we have to devise structures to cope with a variety of situations in which the adjustment process is anything but automatic or easy if we want to smooth the way for innovation.

The structures include the type of company responsibility and communications discussed above, but that is not enough. Some needs of displaced workers are beyond what can reasonably be considered the responsibility of the company. Training or retraining for jobs in other industries and assistance in job search and relocation are really societal responsibilities.

The costs of technological innovation may also extend beyond the plant or company in which the innovation is made. They extend to the community where the effects of a reduced workforce or the extreme case of an older plant being replaced by a technologically advanced one in another location may cause secondary unemployment. The community may need assistance in attracting new industry or in adjustment and relocation programs for the displaced workers.

The effects may also be felt in other companies far removed from the one making the innovation. Rapid or revolutionary innovation in one or a few companies can cause job loss from failure or inability to innovate in other companies. The graphics arts industry is a good example of this. The new electronic equipment can be afforded only by very large companies, and the small private plants and job shops are dropping out of existence.[14] When this happens, there is little chance for the workers to bargain over the effects of the innovation; it is simply something that happens to them.

Finally, a great deal of technological innovation is taking place in nonunionized sectors of the economy. Employers in these companies may act paternalistically and have good programs that consider the needs of their workers as innovations are introduced, but they may not. Workers in these industries are just as much if not more exposed to the effects of innovation on their jobs. They too need a floor of human protections, even though management is not obligated to bargain with them in order to put innovations into place.

There are a minimal set of protections that labor considers important enough to be embodied into our laws. These protections would apply to workers displaced by industrial innovation as well as for other reasons such as business closings, corporate relocation decisions, increased imports, or other economic reasons. While they are being discussed in various forms these days, essentially they are:

1. Early warning or prenotification of impending change.
2. Compensation at least up to the level specified in the Trade Adjustment Assistance Act which is 70 percent of wages for 52 weeks. Longer provisions may be necessary for older workers.
3. Protection of benefits—continuation of health and life insurance and pension for reasonable period of time or until new employment is found to prevent a cutoff of those vital benefits; assistance with mortgage payments during the period of unemployment.
4. Training and retraining for those who need it in order to find new jobs.
5. Relocation assistance.[15]

These or similar provisions are before Congress in various forms in the extreme cases of plant closings of major reductions in force.[16] But there is no particular reason why they should be confined to the extreme cases. Displaced workers of all types suffer economically, socially, and even physically.[17] If their needs are not met voluntarily—and in many cases they are not or cannot be—then there should be basic legislation to require that they are met.

There is an open question here of who should be responsible for meeting those needs. In the case where an ongoing enterprise introduces an industrial innovation that will increase productivity, profitability, and sales, then it is clear that meeting a good portion of the human needs of that enterprise's employees should be a company responsibility—part of the cost of the investment from which the company expects to reap benefits. Society has begun to accept the fact that workers should no longer bear the costs of industrially induced diseases and that communities should no longer bear the costs of industrially induced pollution—that these are costs of production to the company and must be internalized—so we need the same type of realization with respect to technological and industrial innovation.

There is a point, however, at which putting too much of the burden of the adjustment process on the innovating company will impede the introduction of beneficial innovations. The society as well as the private company benefits from the innovation; those efforts that have been made to measure the social rate of return on innovation have shown it to be quite high.[18] Because of this society must bear some of the human costs of innovation.

The public price tag for the type of comprehensive adjustment program just outlined will be quite high, even if companies shoulder that part of the responsibility that is theirs. The experience with the modest program of Trade Adjustment Assistance has been one of difficulty in getting even that fully funded. Movement to a comprehensive adjustment program that could help smooth the way for innovation would be worse than useless without adequate appropriations. Ideally, broadly representative groups of labor and management should sit down together and work out a mutually acceptable division of responsibility for the human costs of innovation between companies and society. That position could then be presented to Congress as a jointly and broadly backed measure. With joint agreement, the probability of passage and full funding would be greatly enhanced. Whether or not such cooperation is possible, however, the corporate world must recognize that legislated protections are becoming an increasing priority of labor. The existence of such a floor of protections to an innovating company in reducing problems would be of great value, and it should be recognized and supported as such.

It is disturbing to note that some companies not only resist their own responsibilities for the human cost of technological change, but also the government assumption of those costs. Assistant Secretary of Labor during the Carter administration, William Hobgood, recently noted the difficulties that exist with the tripartite effort to solve the problems of the auto industry. He pointed out that most auto firms do not share the United Auto Workers' interest in federal assistance for retraining or relocating workers because they do not want to disturb the available labor pool should they resume operations.[19] To a worker, that sounds reminiscent of the conditions John Steinbeck found in the California fields, when the growers advertised for two, three, or ten times the number of workers needed to be sure that the migrants were lean and hungry and ready to work under any conditions offered. The union can see that the auto industry must undergo some changes if it is to be competitive with imports, and the direction of many of those changes—increased use of robots, increased use of electronic components that are purchased from firms in Silicon Valley or elsewhere rather than having wiring for the same function being done on the assembly line, and so forth—indicates a reduced workforce. The union sees that some of their members will be better off being retrained and moving, for example, to where there is a shortage of workers to manufacture the robots that cannot be produced fast enough for industry today or being retrained to program and operate the robots.

A national policy is the way out of this type of dilemma. That policy should grow out of national discussions about the respective roles of companies and society in dealing with the human costs of innovation and should lead to the codifying of the respective responsibilities of each into law. The rules of the game will then be set, the costs and benefits equitably allocated, and the business of revitalizing the economy through innovation can proceed.

Society

All the social mechanisms to smooth the path of labor's acceptance of innovation require goodwill on all sides to accomplish, even if rights and responsibilities are fixed by legislation. It is on that dimension that a troublesome trend is developing in the social-political arena—the fourth level on which we have said that the inventive social structures supporting technological innovation are important. That trend is antiunionism. When corporations in their social-political entities such as the Chamber of Commerce or the Business Roundtable launch the type of bitter attack that they have against Labor Law Reform, or perhaps even more important, conduct a continual sniping campaign against occupational health and safety protections, there will certainly be repercussions when they try to enlist worker support to implement innovative changes. Survey evidence, for example, shows that workers have become increasingly aware of the hazards that exist in their working environments.[20] It is not an overstatement to say that workers perceive attacks on OSHA as stemming from a basic disregard of their most human needs, and that such attacks therefore create a situation in which corporations are perceived as willing to, as it is often extremely stated, "throw workers into the scrapheap." If there is to be the good labor-management cooperation at the plant or industry level that nearly everyone agrees is necessary for smooth adjustment to technological innovation, the level of this type of interference in open and trusting communication will have to be lessened.

The Future

Labor is interested in a healthy rate of innovation in the United States and in the fruits of innovation: increased productivity, competitive industries, an improved standard of living, and lower inflation. Given the types of social structures to support the human costs of innovation discussed here, labor can and will respond positively to specific changes. But labor also has questions about the directions that innovation is taking and about the manning of those directions for the structure of employment in the country. Every look at the future of employment begins with a discussion of the shift from a goods-producing to a service-producing economy; from a blue-collar to a white-collar workforce. Labor is troubled by the implications of that shift. Eli Ginzberg of Columbia University has pointed out that two-thirds of the jobs created since 1950 can be classified as low paying, high turnover, or part time.[21] If a worker is displaced from a manufacturing to a service sector job, it will almost always be with a substantial cut in pay.

There is serious concern over a polarization of the workforce and society. Will the future hold good paying jobs for a few sophisticated technology creators and installers at the top and very low paying jobs for the service providers at the bottom, with very little middle and virtually no way to move from one to the other? Do we need consciously to create a middle or at least a bridge between the extremes? We already have an unacceptably large pool of "structurally unemployed" people. We cannot afford to add the great middle of industrial workers to their ranks.

The mechanisms for peaceful adjustment to technological innovation discussed here assume that adjustment, rather than permanent displacement, is possible, and that even if it requires retraining and relocation, comparable level jobs are available somewhere in the economy for those who are displaced. If that becomes an unreasonable expectation, workers will naturally cling to what they have and try to block change.

We have to bend the direction of technological change to preserve or create new middle jobs. We have to take care to preserve our manufacturing base, even if it is changed in the content of its jobs. In this respect, labor is concerned about the rate at which new technology is exported and the attitude of many people who are willing to accept a service economy role for the United States. It is also concerned about unnecessary job displacement. Perhaps, as Harvey Brooks points out, we should consciously channel innovation to the saving of materials and energy rather than to reducing labor inputs.[22] We also may have to reach back into our educational system to prepare people for jobs in which they must increasingly interact with expensive electronic equipment. The generation of teachers now in the schools did not grow up in a computerized world, and they often have fears of it which are transmitted to the children. Yet every worker and citizen in the next generation should not only have the skills to use the new technologies based on electronics, but also to understand them, program them, and to some extent even create and repair them. If people are acquiring these and related skills in their elementary schools and high schools, if the state of the art is taught early and universally, then it will be much easier to structure the jobs of the future into interesting and well paid combinations of duties.

Finally, we may have to recognize the legitimacy of allowing workers one or two or three periods in their lives during which they will go back and learn new skills. As innovations in medical science and technology extend the possible working life of people, it may be a good investment for society to create an entitlement to one or more paid sabbaticals for each worker during the course of his or her life for that purpose. Educational systems will have to be structured to support retraining and change. This is not a new idea or ideal, but it is one that is far from realization. Labor organizations could play an important role in sponsoring and conducting this periodic reeducation.

Summary

To summarize, labor recognizes the benefits that technological innovation and technological change can bring to the economy and society. Labor is ready to be a willing and even enthusiastic participant, but it cannot allow individual workers to bear the cost of that innovation in human suffering. The mechanisms necessary to elicit a positive response from labor are rather well known; they boil down to a concern and provision for the basic human needs of those who may be displaced. To some extent, companies must be prepared to pay the human as well as the technological and capital costs of innovation, and society must reward them for doing so. But society, through government, must also absorb a fair share of those costs. National discussion and agreement is needed on just how those costs ought to be shared, but they must not be allowed to fall between the cracks. Finally, in pursuing technological innovation, we must take care that our educational system keeps up with the needs of the economy, and that we are retaining or creating a range of opportunities for meaningful and well-paid work and for social mobility. That which needs to be done has been identified and known for some time. A great deal of goodwill and creativity will be necessary to make it a reality.

Notes

1. Louis Edward Alfeld, "Joint Labor-Management Programs in the Construction Industry," National Center for Productivity and Quality of Working Life, September 1978.
2. "U.S. Firms, Worried by Productivity Lag, Copy Japan in Seeking Employees' Advice," *The Wall Street Journal*, February 21, 1980, p. 48.
3. David B. McLaughlin, "The Impact of Labor Unions on the Rate and Direction of Technological Innovation," Institute of Labor and Industrial Relations, The University of Michigan, Wayne State University. Report prepared for the National Science Foundation, February 1979.
4. "AFL-CIO Pension Fund Investment Study" prepared for the Executive Council Committee on Investment of Union Pension Funds by Ruttenberg, Friedman, Kilgallon, Gutchess and Associates, Inc., Washington, D.C., August 1980 and Recommendations of the Committee to the Executive Council adopted August 21, 1980.
5. William A. Schroeder, "The Effects of Changing Technology in Printing," in *Silicon, Satellites and Robots*, Department for Professional Employees, AFL-CIO, September 1979.
6. *Silicon, Satellites and Robots*: The Impacts of Technological Change on the Workplace, Dennis Chamot and Joan M. Baggett, ed., Proceedings of a Conference held in June 1979. Department for Professional Employees, AFL-CIO, September 1979.
7. "New Firm Seeks Shipping Subsidy," Journal of Commerce, July 11, 1980.
8. McLaughlin, "The Impact of Labor Unions on the Rate and Direction of Technological Innovation."
9. Conversation with Phil Ray, Food Industry Joint Labor-Management Committee, Washington, D.C.
10. See for example, Richard E. Walton, "Work Innovations in the United States," *Harvard Business Review*, July-August 1979, p. 97 and Statement by Markley Roberts, Economist, AFL-CIO before the Senate Committee on Labor and Human Resources, September 17, 1980.
11. For example, this became a strike issue between BRAC and the Norfolk and Western Railroad in 1978. Fred J. Kroll, "Bargaining for Change on the Railroads" in *Silicon, Satellites and Robots*.
12. Quoted in *Silcon, Satellites and Robots*.
13. Statement of Carroll H. Teague before the Senate Committee on Labor and Human Resources, September 17, 1980.
14. Schroeder, "The Effects of Changing Technology in Printing."
15. See, for example, the separate testimonies of Howard Samuel, President, Industrial Union Department, AFL-CIO and Markley Roberts, Economist, AFL-CIO before the Senate Committee on Labor and Human Resources, hearing on "The Role of the Workers in the Evolving Economy of the 80's," September 17, 1980.

16. In the Senate by Harrison Williams and Donald Riegle and in the House by William Ford.

17. See, for example, National Advisory Committee on Economic Opportunity, Critical Choices for the 1980s (Washington: U.S. Government Printing Office, 1980) and Arthur Shoestak, ''The Human Cost of Plant Closings,'' *AFL-CIO American Federationist*, August 1980.

18. Edwin Mansfield, ''The Economics of Innovation,'' reproduced in *Seminar before the Committee on Science and Technology, U.S. House of Representatives*, June 18, 1980, (Washington: U.S. Government Printing Office, 1980).

19. Quoted in ''Daily Labor Report,'' October 6, 1980, (Washington: The Bureau of National Affairs, Inc., 1980).

20. Robert P. Quinn and Graham L. Staines, *The 1977 Quality of Employment Survey* (Ann Arbor: Survey Research Center, 1978).

21. Quoted in William Lucy, ''Can We Find Good Jobs in a Service Economy?'' in *Working in the Twenty-First Century*, C. Stewart Sheppard and Donald C. Carroll, eds. (New York: John Wiley & Sons, 1980).

22. Harvey Brooks, ''Technology, Evolution, and Purpose,'' *Daedalus*, Winter, 1980.

chapter six

GOVERNMENT AND INNOVATION
Daniel De Simone

For as long as government has influenced technological innovation in America, and that has been since the beginning of the Republic, there has been a continuing debate about its proper role.[1] The debate centers principally upon government interventions that affect competition and resource allocation in the market economy, but goes beyond that to the broader question of whether the increasing centralization of government planning and decision making is not antithetical to the overall good of society, both materially and politically. Materially, because the federal government has amply demonstrated that it is incapable of efficiently manipulating the complex maze of levers and switches that determine the rate and direction of economic activity (the Department of Energy is but one example); politically, because excessive organization conflicts with individual freedom and creative enterprise.[2]

It is one of the ironies of this debate that the proponents of appropriate technology or "small is beautiful" do not appreciate that the means for achieving the decentralization which they seek is even greater centralization of authority in Washington. And on the international scene, the socialist states, particularly the Soviet Union, demonstrate the poverty of imagination and the paucity of material benefits that result when government planners and controllers make all of the key economic decisions. In the long run, if we are allowed that, the best weapon of the Free World against the expansionism of the USSR may well be its economic system. It is the quintessential example of what governments should not do to promote technological innovation in the civilian sector.

Public policy affects technological innovation in many ways, some of them undisputed, as in the support of basic research and education and the national system of measurement and basic standards. The administration of the patent system is uniquely a governmental responsibility.[3] Other examples are where the government itself is the customer for technology, as in the procurement of weapons systems or exploring the universe. In these examples, the linkage between public policy and science and technology is direct, positive, and commonly understood.

But there are other public policies that affect innovation whose effects are mainly indirect, subtle, and not commonly understood, unintended as well as intended, negative as well as positive—and profoundly important, for they determine the climate for innovation in America. These policies are the ones that lie at the heart of the debate regarding government's proper role in the stimulus of innovation or the shackling of the technological Prometheus. In this category of policies are tax rates, credits, and depreciation allowances that influence investment and risk-taking; monetary policies that affect the money supply and the availability of venture capital; the indiscriminate enforcement of the antitrust laws on the one hand and the rescue of comatose corporations on the other; and the burgeoning accumulation of regulations that impinge upon every step of the innovation process. To list these policies is not to pass judgment upon their net worth, but merely to recognize that they are the most powerful forces on the field of battle of the innovation contest.

Economists differ on the extent to which innovation contributes to economic growth and productivity.[4] It is questionable whether this contribution can be measured with precision. In any case, the palpable realities are that innovation is indispensable to domestic vitality, social progress, a vigorous world trade posture, and to the security and prosperity of the Free World. The industrial democracies have sought to spur research and development and the applications of technology for these ends. In the United States, one way or another, all of the instruments of government have been involved: executive, legislative, judicial, and regulatory. But the record does not speak well of their collective impact on the climate for innovation in America, for the indicators show that this climate has faltered and declined since the 1960s.[5]

During the past two decades and over the course of five presidential administrations, federal policies and programs influencing technological innovation have been reviewed and debated periodically in a quadrennial display of concern for America's technological capabilities. In the Congress, scores of legislative proposals to stimulate invention and innovation have consumed the energies of committees and study groups. And in the courts, questions of profound significance for industrial innovation and competition have been and continue to be litigated.

In all three branches of government, such questions loom larger now than in the decades past. The 1980s will be a watershed in shaping the climate for innovation in the United States. Whether this climate will be salubrious and socially beneficial or stifling and counterproductive will depend upon the wisdom and understanding that are brought to bear on questions of policy affecting innovation and entrepreneurship in America.

The 1980s will be a critical period in which we will be contending with several new realities, including rising world demand for scarcer and costlier resources and a rapidly changing world division of labor and markets. Merely to do the same with less (the so-called "constant pie") will not be good enough to meet rising domestic aspirations and world responsibilities. I agree with Harvey Brooks that those who argue that economic growth should be arrested and the problem of poverty solved by redistribution alone cannot be taken seriously.[6] Accordingly, if we are to do more with less, through new capabilities and higher rates of productivity, the pace and direction of innovation will be a key factor. The margin of error afforded by the materials and energy abundancies of the past is no longer available to bail us out after failures of policy and will. As never before, social and technological innovation will be required for economic growth and the creative renewal of our society.

The record of the past decade regarding the innovative health of the United States has never received more attention in the press than it has over the past year. Every major newspaper or periodical has published feature articles on the decline in U.S. productivity, international competitiveness, and entrepreneurial initiative. The various measures of this decline—the "innovation indicators"—are sobering.[7] To be sure, they are necessarily imprecise and can even be misleading if misread. For example, if we consider absolute figures and not just rates of change, the United States is still ahead of the rest of the world in productivity and investments in research and development. Moreover, productivity rates vary from industry to industry, as do competitive and technological capabilities. Ratios of research and development expenditures as a percentage of GNP are debatable indicia of cause and effect relative to *innovative* performance. And patent statistics can be misleading harbingers, too. They lump together patents on inventions that range in novelty and significance from marginal improvement designs that are barely patentable to the relatively few inventions that lead to radical breakthroughs in technology.

One indicator is spending on research and development (R&D). As a percent of Gross National Product, R&D funding, both public and private, was down by about 20 percent in 1978 from what it was in 1968.

A second indicator is drawn from the first one: the amount of R&D spending that goes to basic research. It has been argued that the fruits of basic research are universal commodities which are not appropriable to the United States alone and therefore cannot validly be employed as an index of U.S. innovation capability. A corollary of this argument is that basic research, by definition, has no practical application. True, but the practical outcomes of science are unpredictable. No one, for example could have foreseen the monumental technologies that were spawned by Maxwell's equations or the awesome consequences of the early research in quantum physics. And if the reservoir of science is for everyone's use, it nevertheless bespeaks a creative energy in the nations that contribute to it. It is a vital sign, not the social pulse rate or blood pressure (the practical measures of civilization), but the vision and sense of value and purpose of a nation. It tells a lot about a society's attitude toward the future, for it is a creative and optimistic country that invests for future generations.

What does the record show about the U.S. commitment to basic research? The fraction of GNP devoted to basic research was about 20 percent less in 1978 than it was in 1968. In adjusted dollars, however, the support for basic research is about the same now as it was then.

A third indicator is private investment in R&D. This is the index most closely associated with real economic growth and the ability to compete internationally. From this investment comes the technology to create most of the permanent new jobs. The United States is now spending less in this column, as a percent of GNP, than either Germany or Japan.

A fourth indicator is the number of patents granted here and abroad to the United States and its major trading partners. The United States granted 9,567 patents to non-Americans in 1966, while 45,633 patents were issued to American inventors, a ratio of almost 5 to 1 in favor of the United States. Since then the patent balance has shifted dramatically. The number of foreign patents granted to Americans dropped to 33,181 over the 10-year period ending in 1976, while the number of U.S. patents issued to foreigners had almost doubled, to 18,744.

A fifth indicator is corporate profitability, and the key measurement here is not gross dollars, but the actual purchasing power of the dollars left after taxes. After squeezing out the inflation and putting aside enough money to cover replacement at today's prices for the plants and equipment that are wearing out, there is less left to encourage new ventures, plants, and equipment now than there was 10 years ago.

A sixth indicator is the burden being placed on the nation's productive apparatus by government. Given a capital-starved economy and double-digit inflation, managements will tend to avoid the longer-term risks for those that are shorter-term and less uncertain. Government regulations have mushroomed over the past decade; so, too, have the perceived risks for investments in innovation. If this view of regulation as an

excessive burden beyond necessity was not shared in the early 1970s
by the universities, it most assuredly is now.[8] It seems as much resented
by university researchers as it is by unemployed steel and auto workers,
although the impacts can hardly be equated. In the one case it is a
question of interference and principle; in the other, a matter of basic
necessity, dignity, and dismay for workers and their families.

In summary, none of these indicators by itself gives a reliable cause-
and-effect forecast of innovative vitality, but in the aggregate they give
a sobering picture of change in the climate for innovation since the late
1960s. The trends mean more than the figures.

The point merits emphasis that R&D statistics tell us something, but
they can also mislead. The general misunderstanding of the significance
of technological innovation, both in terms of its potential and its requirements, arises from the tendency to equate it with research and
development. This pitfall was highlighted almost a decade and a half
ago in the report of the Charpie Panel.[9] In fact, R&D is but a fraction
of the effort, investment, and risk-taking required to bring an invention
to market as a full-fledged commercial product or process. In successful
product innovations, for example, the percentage of costs attributed to
R&D is typically less than 10 percent, which is to say that a risk-taking
entrepreneur must contemplate 90 percent of his costs *after* the R&D
stage. In this context, one can appreciate the inadequacy of the Carter
Administration's "innovation" policy proposals, most of which are
aimed at R&D, not the more costly risks of innovation.[10]

The record of innovation policy proposals in the previous presidential
administrations has also been mixed. There is definitely bipartisanship
in this record. There have been high points, but many of the recommendations have ranked with the Carter proposals in terms of feeble-
ness of purpose and impact and are illustrative of a kind of Gresham's
Law that pre-empts effective action: bad innovation policy drives out
the good. It would be more charitable to say that policies that purport
to address innovation but are, in fact, aimed at inconsequential aspects
of R&D, *pre-empt* effective action on the innovation front. Let it also be
noted that this is not meant to be a criticism of the Science Advisers to
the President, of which the nation has been blessed with an extraordinarily capable succession of gifted individuals. For the institution of the
Executive Office of the President, and the White House apparatus, with
its competing forces like so many dukes in a regal court, including the
Office of Management and Budget as chancellor of the exchequer, is
one in which science advice is lucky to survive let alone prosper.

Each of the past five administrations has examined the climate for
innovation and found it wanting. The Kennedy administration made
the first attempt to establish a uniform government patent policy,

pushed tax proposals as indirect stimuli for innovation, launched a Civilian Industrial Technology Program to develop ''generic'' technologies for industries with low rates of R&D, and established projects to aid inventors and transfer knowhow to small businesses. The Johnson administration established the State Technical Services Program (an industrial analog of the Agricultural Extension Service) and made recommendations concerning taxation, venture capital, antitrust, and patents, all of which were aimed at encouraging the establishment of new, technologically based enterprises. Like so many aspects of national life at that time, these recommendations were a casualty of the Vietnam War and the ''guns *and* butter'' inflation that was beginning to smoulder.

The Nixon administration undertook a massive review in its New Technology Opportunities Program, which resulted principally in three programs of study and experimental R&D incentives, as well as the first presidential message to Congress on science and technology. As Vietnam was the headsman of the Johnson era, Watergate took its toll here.

The Ford administration's review resulted in a draft report, ''U.S. Technology Policy,'' which was released at the beginning of the Carter administration and which recognized the importance of technological innovation for national purposes and recommended easing the burdens on technological entrepreneurs. Finally, the Carter administration, in an intensive 18-month study reminiscent of President Nixon's New Technology Opportunities Program, came up with similar results and recommendations, as well as a presidential message to Congress.

Clearly, there has been no lack of attention in the executive branch over the past twenty years to measures specifically targeted at research and development. Of greater importance and impact, however, were the regulatory and economic policies that indirectly affected technological innovation in immeasurable, but nevertheless profound ways. The recurrent litany of innovation policy proposals over the past two decades persuades me that we are missing the mark. They have been mostly panaceas cast in terms of impending national ruin, vaguely defined as the eclipse of the United States in world economic and political affairs.

For a time, I thought of these nostrums as first principles. It is a mark of some civilizing influence, for which I am grateful, that I no longer regard them as such. I do not think the nation would be ruined if all of the current legislative proposals specifically targeted to spur technological innovation were jettisoned. I believe, with Montesquieu, that if indeed it be true that the United States in on the brink of ruin should these proposals not be signed into law, that would merely confirm that there is an underlying cause at work which would prevail in any case. And that brings me to what I perceive to be that underlying cause.

It is my belief that the erosion of innovation and venturesomeness in

America has resulted more from the failures of economic policy than from the absence of effective measures directly targeted at research and development. Certainly, this erosion is not due to any deficiency in the capabilities of U.S. scientists and engineers, who are the envy of the world, or to the level of R&D expenditures. Nor is it attributable to any shortage of creative ideas and inventions. The problems lie in the further steps that are required to bring inventions to reality, i.e., to carry them all the way through the innovation process. Policies that narrowmindedly and without redeeming virtue impede these further steps and discourage risk-taking over the long term stifle entrepreneurial spirit and discourage innovation.

On the congressional side, the activities of the last two decades are equally instructive. In a variety of ways, Congress has attempted to improve the climate for innovation. In other ways, it has inadvertently, or sometimes deliberately, erected barriers to industrial innovation or attempted to prescribe it, particularly in the context of health, environmental, and safety regulations. In still other respects, congressional (and presidential) initiatives have pre-empted scientific and technical manpower for federal mission-oriented programs, especially in the fields of space and defense and increasingly in recent years on subsidized energy projects such as solar, gasohol, and synfuels. The supply of such talent is relatively inelastic in the short term, and, therefore, irrespective of the merits of each particular case, applying this talent to federal missions necessarily denies it to other sectors of the economy.[11]

I am not here questioning the priority that is being given to defense and to so-called energy "independence." The trade-offs, however, should be understood. Conversely, an even-handed assessment must consider the fiscal consequences of policies to spur innovation and productivity on the budgetary requirements of national defense. Policies to promote growth and renewal—and, therefore, to support defense in the future could be competitive with the needs of defense in the present.[12]

The Space Program is a special case. Congress lavished attention on it during the 1960s, and it grew from a shaky start to a magnitude that transported Americans to the moon. It also made communication satellites a reality and stretched the capabilities of exploration and knowledge of the universe by leaps and bounds unmatched since Copernicus and Galileo stunned the world with their revelations of the earth's proper station in the cosmos. Although there were minuses (opportunity costs) along with pluses for industrial innovation in these space initiatives, what was said earlier about the worth of basic research applies to space exploration as well.

Meanwhile, Congress killed the supersonic transport and the antiballistic missile system in debates of such acrimony and emotionalism that their effects are still felt in the continuing reviews of strategic arms issues and the future of the U.S. aircraft industry. Many other examples could be listed of the ways in which Congress directly affected technological innovation during the period, to say nothing of the telling impact of economic policies on the willingness of entrepreneurs to take the long-term risks that technological innovation requires. What seems to be lacking in general on Capital Hill and indeed throughout government is an understanding and appreciation of how technological innovation is spawned, nurtured, financed, and transformed into new technological enterprises that grow, provide jobs, invigorate society, and satisfy the needs of people.[13]

Two technologies, computers and telecommunications, began to reach revolutionary status during the period, and Congress has been grappling with them, so far without success, particularly in its efforts to rewrite the Communications Act of 1934. Both of these fields of technology have been propelled by startling advances in microelectronics, and further advances are in the offing through very large-scale integrated (VLSI) circuits and fiber optics. Another portentous technology emerged during the period: biotechnology or "genetic engineering," which some predict will have as great an impact on industry in the twenty-first century as chemistry and physics did in the twentieth. Whether this occurs sooner or later, biotechnology, like computers and telecommunications, may well have societal impacts no less consequential than those of the Industrial Revolution, although different.[14]

In the courts, these revolutionary technologies are central to landmark antitrust and patent cases that could have a dramatic impact on industrial innovation. Just recently, the U.S. Supreme Court handed down a split decision on the patentability of bio-technologies and resultant life forms, ruling that the Patent Law does not preclude the granting of patents on genetically engineered life forms. Far from being settled, the controversy attending this case has hardly begun, for the Court has, in effect, passed on to the Congress the question of whether or not life forms *should* be patentable and, if so, where the line would be drawn as the geneticists go up the life chain.[15]

In the antitrust arena, the courts are struggling with the complexities of telecommunications and computers. The United States is an acknowledged world leader in these fields. AT&T and IBM, the firms most responsible for this pre-eminence, have been defendants in the biggest antitrust suits ever brought by the U.S. government.[16] The kinds of technical questions that were contested in these suits are perhaps the most complex in the field of jurisprudence. In related private antitrust actions, juries have admittedly been dumbfounded by the web of technologically laden issues argued before them. It is questionable whether the judicial system can cope effectively with such issues in jury trials.

It is also questionable whether conflicts of law and policy can be satisfactorily resolved in such contests in a way that enhances both competition and industrial innovation.

These are merely highlights of the impacts on innovation of the legislative, judicial, regulatory, and executive apparatuses of government. The policies they establish and pursue are more often in conflict than in harmony. Sometimes it would appear as though they were giant wrecking machines flailing at one another across the Mall in Washington. But out of this, one way or another, social choices emerge. The question is whether they are consciously made with an appreciation of their consequences.

Despite all of the rhetoric of the presidential campaign of 1980, the United States is the foremost economic and military power in the world. Lodged in this pre-eminence, however, is an Achilles' heel, for success breeds complacency. It is the bane of the successful—and nation states are no exceptions—to regard success as a terminus instead of a step in the interminable process of evolution. Natural selection applies indiscriminately in a theoretically pure market economy, but even in the modified free market, government props and other interventions cannot void the winds of change. At some level the impact will be felt. If the buffeting is deflected to spare an ailing or dissipated firm or industry, it will be felt eventually at the level of the state.[17]

This is not to suggest that governments should stand aside as mere witnesses of battles for industrial survival or commercial plunder. Nor on the world scale is it to justify in the name of inevitability the beggaring of the developing world. Rather, it is to underscore the importance of understanding the trade-offs when governments decide to intervene, which requires that the overall impacts be understood when choices are made: If we do this, what is the price? If we do not do this, what are the likely consequences? What else might we do and what would that entail?

Social Darwinians will disagree, but it is the social unacceptability of certain consequences of natural selection (which, again, will occur at some level; if not the firm, then the state) that demands social invention. At its best, social invention is based on rational choice; at its worst, it emerges from aimless or passionate stabs in the dark. It is regrettable that public policy has tended more toward the latter than the former during the 1970s.

I would suggest a law of "Conservation of Social Cost". Social cost can be neither created nor destroyed through social invention, which simply redistributes it. How it is redistributed is the central challenge to society. The question is: is this to be done with single focus on single interests or with care for the long-term direct and indirect effects on all factions and on society as a whole?

Proposed social choices to cut back or curtail economic growth are the most troublesome, for they exalt environmental values over other ones that are strongly felt, not just by business, but emphatically by spokesmen for the disadvantaged.[18] Bayard Rustin typifies this view:

> It has become part of the conventional wisdom of certain segments of the intelligentsia that economic growth must be significantly reduced or halted altogether by the wealthier industrial countries. The advocates of the zero-growth philosophy more often than not are passionate supporters of a more equitable social order, particularly for racial minorities here and for the masses of the underdeveloped world. That there is a glaring contradiction between their opposition to growth, on the one hand, and their avowed support for a more equal society on the other, seems not to have dawned on many of these people. In their naive idealism (and, in some cases, scarcely camouflaged elitism), they would deprive society of a key means for ending poverty and misery at the very moment that the most deprived groups are demanding a rightful share of the fruits of the modern industrial, technological, economic order.[19]

It is not surprising that this clash of values will occur. People out of work, because of what they feel, rightly or wrongly, has been excessive concern for environmental purity, will regard ardent environmentalists as "elitists" who display "exquisite sensibilities and moral virtue," but fail to appreciate that "poverty is degradation, misery and starvation, not the level of carbon dioxide in the air. Growth is the best solution to poverty."[20]

In a democratic society such conflicts can and should be worked out. It is regrettable when they are pushed to an "either-or" impasse, for we can have both employment *and* environmental protection, within reason. The messianic zeal to perfect the environment at all costs is what violates equity and reason.

We all know that some of the most ingenious technologies can go awry or be misused. The only safety that can be hoped for in this is the intelligent exercise of day-to-day judgment. Turning off nuclear power plants means turning on coal-fired ones. People, particularly those at the bottom of the economic ladder, are not going to wait for the day, beyond prediction, when an abundance of solar-electric energy becomes available. The risks of nuclear energy may be traded for the risks of coal, but they will not go away. Learning how to cope with risks in a rational way is perhaps the most fundamental challenge to a free society.

Crusades to turn off technological advance, rather than channeling it along beneficial lines, are futile and can lead only to frustration. Like it or not, technology is indispensable to world peace and prosperity. Consider that each day adds a new Toledo, Ohio, to the world. That amounts to 250,000 human beings each day, or 90,000,000 people a year as these "cities-per-day" keep on adding up.[21] It is technology that has made the support of these staggering increases possible, and it is the increasing aspirations of these multitudes that make further advances in technology necessary. True, there is a dilemma here, but what would the zero-growth advocates suggest we do with these human beings pending effective and acceptable population control?

It is easy to lament the problems that have resulted from some technological applications and to advocate a return to the simpler, seemingly pastoral lifestyles of earlier times. I doubt that there would be many who would book passage. "No one in his senses would choose to be born in a previous age, unless he could be certain he would have been born into a prosperous family, that he would have enjoyed extremely good health, and that he could have accepted stoically the death of a majority of his children."[22]

We have no choice but to go forward, striving to increase our understanding of the world and the universe, and making the best of each step of the way.

In conclusion, an observation about the human race, which is refreshingly American and timeless:

> If I am right, it will be a slow business for our people to reach rational views, assuming that we are allowed to work peaceably to that end. . . .
> If I fear that we are running through the world's resources at a pace we cannot keep, I do not lose my hopes. I do not pin my dreams for the future to my country or even to my race. I think it probably that civilization somehow will last as long as I care to look ahead—perhaps with smaller numbers, but perhaps also bred to greatness and splendor by science. I think it not improbable that man, like the grub that prepares a chamber for the winged thing it never has seen but is to be—that man may have cosmic destinies that he does not understand. And so, beyond the vision of battling races and an impoverished earth I catch a dreaming glimpse of peace.
>
> Justice Oliver Wendell Holmes, Jr.[23]

Notes

1. See, for example, Alexander Hamilton, "Report on the Subject of Manufactures," Section VIII, submitted to the Congress, December, 1791; reprinted in J. E. Cooke (ed.), *The Reports of Alexander Hamilton* (New York: Harper & Row, 1964), p. 175.
2. See, for example, Senator Paul H. Douglas, J. Enoch Powell, *How Big Should Government Be?* (Washington: The American Enterprise Institute, 1968); Friedrich A. Hayek, *The Road to Serfdom* (Chicago: University of Chicago Press, 1944).

3. The only public policy directed towards science and technology that is specified in the Constitution of the United States: Article I, Section B.

4. See, for example, John W. Kendrick, "Productivity Trends and the Recent Slowdown: Historical Perspective, Causal Factors, and Policy Options," *Contemporary Economic Problems* (Washington: The American Enterprise Institute, 1979): Edward F. Dennison, "Explanations of Declining Productivity Growth," *Survey of Current Business* **59** (August 1979): 1–24.

5. See, for example, Mary Ellen Mogee, "Technology and Trade: Some Indicators of the State of U.S. Industrial Innovation," 2nd Session, April 21, 1980; *Science Indicators 1978* (Washington: National Science Board, National Science Foundation, 1979); Irving Shapiro, "Technology's Decline: America's Self-Made Paradox," speech to the Economic Club of Detroit, January 22, 1979; Richard Atkinson, "Reflections on Science in American Society", speech at Stanford University, March 14, 1979.

6. Harvey Brooks, "Technology, Evolution, and Purpose," *Daedalus*, Winter 1980, p. 78.

7. Magee, "Technology and Trade."

8. A. Bartlett Giamatti, "Researcher Versus Bureaucrat," *The Washington Post*, October 15, 1980; E. Marshall, "Universities 'Battered' by Federal Regulators," *Science* **202** (1 Dec. 1978): 955.

9. D. De Simone, *Technological Innovation: Its Environment & Management*, Report of the Panel on Invention and Innovation (Washington: U.S. Govt. Printing Office, 1967), p. 9.

10. Fact Sheet, "The President's Industrial Innovation Initiatives," The White House, October 31, 1979; "Industrial Innovation," Joint Hearing before the Senate Committee on Commerce, Science, & Transportation, 96th Congress, 1st Session, October 31, 1979.

11. J. Herbert Hollomon, in W. N. Smith and C. F. Larson (eds.), *Innovation and U.S. Research* (Washington: The American Chemical Society, 1980), p. 197; see also, D. De Simone, ed., *Education for Innovation* (Oxford: Pergamon Press, 1968), p. 23.

12. H. Stein, "Curriculum for Economics 1981," *The AEI Economist*, American Enterprise Institute for Public Policy Research, June 1980, p. 9.

13. D. De Simone, ed., *Education for Innovation* (Oxford: Pergamon Press, 1968).

14. *Facing the Future*, The Interfutures Project (Paris: OECD, 1979), p. 117.

15. Diamond, Commissioner of Patents and Trademarks v. Chakrabarty, Civil No. 79–136, U.S. Supreme Court, June 16, 1980.

16. United States v. American Telephone and Telegraph Co., No. 74–1698 (D.D.C., filed Nov. 20, 1974); United States v. International Business Machines Corp., No. 69–200 (S.D.N.Y., filed Jan. 17, 1969).

17. This is a variation of the biological metaphor discussed in H. Brooks, "Technology, Evolution, and Purpose," *Daedalus*, Winter 1980, pp. 68–70. It is based on the hypothesis that a market economy has its own form of natural selection and that government interventions thwart that process.

18. See, for example, H. Brooks, "Technology, Evolution, and Purpose"; the central question is "innovation for what?"

19. Bayard Rustin in "Small Is Not Beautiful," a speech presented to Americans for Energy Independence, June 1977.

20. W. Beckerman, quoted in *Time*, June 6, 1977, p. 63.

21. R. L. Strout, *The Christian Science Monitor*, October 5, 1977, p. 22.

22. J. H. Plumb, *Technology Review*, July/August 1976, p. 57.

23. From a speech at a dinner of the Harvard Law School Association of New York, February 15, 1913, reprinted as Senate Doc. 1106, 62nd Congress, 3rd Session.

chapter seven

POLITICS, POLITICIANS AND THE FUTURE OF TECHNOLOGY

Clarence J. Brown

Capital investment decisions are both causes of changes in society and occur as a result of those societal changes. Henry Ford's invention of the assembly line techniques that brought America out of the horse and buggy era not only spawned great industries and unions and cities, but it made possible industrial, marketing, and societal changes in other fields as well: the modern highway system and the shopping center.

The industrial unions were social inventions to which federal law was obliged to respond, not only with laws affecting labor's rights and obligations, but with changed tax laws addressed to a new laboring middle class.

And so we should explore this chicken-and-egg interrelationship between technical change which is so vital to national economic progress and the societal changes which cause it and which it causes. But we should also look at the role of government action in shaping or responding to both technical and social invention.

Government actions of all kinds provide incentives and disincentives for certain kinds of social behavior. Each time the government acts, it sets into motion certain incentives or disincentives to social inventiveness which may or may not be consistent with the original purpose of the government action. The rewarding by Congress of veterans of the American Revolution with rights to western lands encouraged settlements of the midwest. More recently, however, we have established a system of welfare and other benefits for the nation's poor which, in many cases, acts as a disincentive for people to enter the labor force. Until recent changes were made, our system of taxation—with its high marginal rates, the double taxation of dividends and the deductibility of interest payments on consumer purchase loans—tilted incentives heavily toward consumption and away from saving. After reassessing these impacts, Congress supported the Administration recommendations to change tax law in the hope of setting different economic reactions. They provide more incentives to work, save, and invest.

Enlightened legislation involving technology (and its economic and social consequences) should consider its impacts in other areas as well. The pressures and constraints inherent in the American political process, however, make it almost impossible to reverse this process and adapt legislation neatly to the social imperatives brought about by technological innovation. These pressures and constraints can be reduced to three observations: (1) politics generally deals with immediate, short-term, direct agendas and not the long-term, indirect future; (2) the art of anticipating issues brought about indirectly by technological change is still too imprecise for politicians; and (3) the political issue often appears larger and more direct than the technological one which gave rise to it or to which it gives rise.

It should be noted that social inventiveness is both politically "neutral" and, in another sense, highly partisan. They are neutral in the sense that some form of social inventiveness takes place in response to most technological development. Sometimes the response is accomplished through private sector actions, and sometimes it is the result of government. The particular form of response, however, can become a highly charged political issue. This is because the form of response often reflects a prevailing political philosophy. For example, more government intervention versus less government intervention is, depending upon which position one takes, going to rule out certain kinds of social inventions.

Public Policy and the Future of Technology: Three Problem Areas

The principles described above, if carried into effect, would require public policymakers to anticipate the effects of decisions more broadly and deliberately than they have in the past. Insofar as possible, the language of a law must reflect future development. Thus, the practicing politician must depend heavily upon the findings of professional forecasters (economists, sociologists, political scientists, and others) who could be loosely described as "futurists." Futurism, of course, is a sophisticated though emerging technology which has already proved its usefulness to policymakers. The term "futurist" describes the sophisticated social scientist as well as the public policy specialist who attempts to explore the ramifications of certain courses of action.

The politician enters the futures business at his or her own peril, however. By definition, politics must deal with the present. To the average constituent, the issues of present-day conditions are a greater influence on his judgment than anticipated future conditions. People who are unemployed, for example, understandably are not interested in whether today's high interest rates will result in a better job opportunity ten years from now.

Thus, legislation usually responds to the perceived *current* needs of society. It is more often reactive than pro-active. To balance this historic practice, political constituencies may have to be more aware of the longer term future, as well as the present, if public policy is to respond effectively to rapid technological development.

A second problem area which policymakers face when trying to look into the future is that the profession of politics imposes greater practical constraints on them than futurism imposes on futurists. Futurists typically chart alternative scenarios within which they expect events to move. Sometimes the scenarios are so different, or the parameters so large, as to be less than useful for making firm public policy decisions. Where should energy policy lead us if predictions about the state of energy technology in the future vary so widely? Should banking policy assume that we are capable of moving smoothly toward a "checkless society" in the foreseeable future? Predictions vary widely!

Partly for this reason, the findings of futurists are often utilized to support or oppose particular points of view which have already been taken by the parties to a political debate. This reflects the fact that the form of response to a given technological development can become a highly partisan issue.

The third problem is related to the one just mentioned. Many social issues have arisen because of technological change and the effect of that change on society. However, the resulting political and social issue may be much larger than the technological one which gave rise to it. Even a clear view of the future (if that is ever really possible) may be of little assistance in coping with the issues of the proper role of government in society and the effective limits of government intervention. For example, the development of television has created a change in the ability of children to choose their own entertainment and the impact of advertising directly on children. Does this mean, however, that the government should regulate the content of advertising on children's TV programs, or is that in conflict with first amendment rights of both programmers and viewers? For a price technology may be capable of creating even more fuel efficiency in automobiles than at present, but does this mean that the government should force stricter standards? Should we also require, in spite of consumer costs, stricter auto safety standards (e.g., air bags) where the technology exists?

Proponents of a government solution might argue that, in the absence of federal action, thousands of people would be adversely affected. But opponents often make the same point that the economics of a free market will be the best arbiter of what should be bought and sold and at what price safety can be afforded, and that the government should not interfere unduly with consumer decisionmaking. The abrupt change in policy which has come about since the Reagan Administration took office turns on issues such as these, and billions of dollars in spending cuts have been predicated upon this change in government philosophy.

Those who differ in their philosophy of government may not differ on what the aims of government should be. Political leaders are not voted into office in order to be apathetic about less safe or more costly automobiles. They may, however, disagree on what government should do about the issue.

Much of the difference turns on the way policymakers perceive the relationship between government actions, private judgment, and economic growth. In a complex economy such as ours, this relationship may be obscure or heavily diluted. But government regulation comes with a price, nonetheless. The price is both personal freedom and economic opportunity. The billions which American corporations have spent in employee safety equipment and pension reform—and filling out the government forms to show they have complied with these requirements—show up in the price of products just as though they were a separate item in the bill.

If the Past is Any Guide, We're in Bad Shape—But it Could be Worse

The landscape is littered with the failures of government policy and the societal response to new technologies. Nevertheless, there have been some success stories where the government has created institutions to cope with the problems created by technological and economic progress.

Too often the government response to a new situation (e.g., finding that air pollution carries health hazards) is to legislate as if the burden of complying with government requirements could be absorbed without cost, or as if the cost were unimportant so long as it can be hidden. And yet almost every government action eventually imposes some different economic cost on some new segment of society. This is illustrated by government's failure to fashion a practical urban policy. The following section examines this phenomenon in greater detail.

Urban "Policy." Depressed inner city neighborhoods and the threat of bankruptcy in a number of large cities attests to the failure of urban policy today. Ever since the McCormick reaper displaced farm workers who eventually became part of our urban labor force, the problems of America's cities have reflected technological invention and innovation and also have brought the need for social invention.

As agricultural production became more efficient and no longer labor intensive, the offspring of previously large farm families were obliged to find factory work opportunities in big cities where they lived within walking distance of work. First public then private transportation made it possible for them to live further and further from the work site. Shopping opportunities remained centralized for a while but also began to decentralize. And then even the factories began to decentralize as they chased the work force and new markets and management and found that communications technology made decentralization more practical.

Some would claim that government has not developed a comprehensive urban policy in response to these changes of the past 50 years. This was the basis for a major initiative to produce one during the last federal administration. In fact, we have had an urban "policy," in that the federal government for decades has been making decisions which affect our nation's cities. These decisions, coupled with social, economic, and technological developments during the post-war years, provide a good example of how these phenomena interact.

On one level this country's urban problem is the result of changes in patterns of production, population distribution and land use. On another level it is the natural result of many unrelated developments whose cumulative effect was not foreseen. Some of these developments were prosaic, such as the development of air conditioning. Others were momentous, such as the enactment of the G.I. bill, but were not intended at the time to affect our cities. It is instructive to list the major postwar effects on our urban areas according to government action, social invention, technological change, and economic development. An abbreviated list would look something like this:

Government Action

- The national highway system, whose original purpose was partly to facilitate inter-city travel, has also facilitated the flight to the suburbs and to the sunbelt.

- The FHA and VA mortgage programs, which have been successful in enabling millions of Americans to own their own homes, has also placed pressure on real estate development outside central cities and urban areas and reduced the tendency to live in row houses, tenaments, or apartments.
- The tax treatment of real estate has encouraged the construction of new, more depreciable plants in outlying areas and the sale, at capital gains rates, of central city property.
- Federal economic development subsidies have enabled many sunbelt communities to construct the infrastructure of sewers, schools, and roads needed to accommodate new industries while keeping taxes relatively low.
- The tax treatment of industrial development bonds has favored the construction of new plants outside urban areas.
- The federal subsidies to air travel through airport facilities grants and design of military aircraft which had subsequent civilian application and government purchased aircraft have not been matched by subsidies to rail travel. This has affected the economy of inner city areas and also encouraged more nationwide travel to the south and southwest.
- Federal policy has encouraged geographic distribution of government procurement centers, including the construction of military facilities in sunbelt areas.
- The G.I. bill brought education and a higher standard of living, which in millions of cases meant new home ownership to a generation of Americans.
- At the state level, state law in many sunbelt jurisdictions enables cities to acquire outlying areas as the metropolitan area expands. Thus, the suburban tax base inures to the benefit of the city as a whole, which is not the case in northern, urbanized states, where cities have grown together but have found it too complicated to jointly resolve their problems or consolidate territorial jurisdictions.

Technological Developments

- Improvements in airplane technology and productivity have minimized the cost and inconvenience of traveling long distances to supervise branches, market to customers, visit family, or take vacations.
- Improvements in automotive technology coupled with the national highway system have made long-distance travel out of the city comfortable and relatively trouble-free.
- The introduction of computer communication has revolutionized

business behavior, enabling large firms to own plants in different parts of the country while maintaining accurate accounting and management systems.

- Air conditioning technology has made living in the sunbelt states more comfortable year-round.

Economic and Social Developments

- A general increase in postwar affluence has encouraged the ownership of second homes and more widespread travel outside the city.
- The continued erosion of geographical and family ties among Americans has helped remove inhibitions against settling in distant parts of the country.
- Suburban shopping centers—an American invention—have changed the urban economic landscape.
- Patterns of racial discrimination and unrest in urban areas have hastened the flight to the suburbs of those who were economically able, thus eroding the inner city tax base.
- The use of tax anticipation notes and sophisticated accounting devices has postponed the day when some cities have had to face up to a deteriorating financial condition.
- High energy costs have encouraged settlement in the warmer areas of the country.
- Modern factory design has made it advisable to construct new plants outside congested urban areas.

This partial list illustrates a number of important points about the appropriate political response to problems brought about by technological and social development.

Most of the items listed above involve second or third order effects on urban areas. The first effects may be beneficial. The harmful side effects may not appear until much further down the line and probably were not foreseen. The immediate effect of the VA loan system, for example, was that millions of veterans had an opportunity to buy their own homes. The second order effect was the employment and materials demand for the construction and furnishing of suburban housing. Many would argue that these beneficial results made the program an unqualified success. It was not until later that the impacts of "urban sprawl" began to be questioned.

Because politics deals with the present and because the second and third order effect cannot always be accurately predicted, the political issue to be resolved usually deals only with the present problem and only rarely tries to anticipate the longer term problems. Should Congress (or could it) have factored in the possible third order effect on the urban economy of the VA loan program or the interstate highway system when it enacted the authorizing legislation? That question implicitly has a negative answer. However, public policy has taken the other side at times. In passing the National Environmental Protection Act, for example, Congress decided that business and government should anticipate every known environmental impact of a contemplated major action. And the Occupational Safety and Health Act says, in effect, that every known or possible safety and health hazard must be considered in the design and operation of work places. These examples illustrate the dangers of attempting to legislate the principle that all known or logically foreseeable effects of public policy should be anticipated: both the National Environmental Protection Act and the Occupational Safety and Health Act are frequently cited examples of needlessly heavy-handed government intrusion into private decisionmaking.

The varied complex of interests affected by social, economic, and technological events often makes the political or government-imposed solution unworkably crude. For example, suburban shopping centers have contributed substantially to the economic weakness of the inner city. But does this mean that the federal government should have stepped in at the possible expense of those who benefit from them and regulated their growth? Should Congress place impediments to free travel on commercial or residential development near our interstate highway system to prevent superhighways from drawing commerce away from urbanized areas? The question is not far-fetched. In my own constituency a leg of the interstate system making a beltway around a metropolitan area was opposed by officials of the central city because they feared competitive development near the highway would detract from other parts of the area. In their view, resolution of congestion problems in the area where the highway would be built was not an adequate justification.

Some believe that the United States should have a national land use policy so that suburban growth is more carefully implemented from Washington. But even recent Congresses, heavily weighted to the "Washington knows best" philosophy, have overwhelmingly rejected legislation calling directly for federal authority for land use planning. While acceptance of local zoning and land use ordinances is widespread and while many of the goals and methods in these ordinances are common, the separation of responsibilities in the federal system has survived well in this case. Land use planning is clearly limited to local responsibility. And yet, the Federal Clean Air Act has become an authority for federal land use planning just as clearly as if it had been called "The Clean Air and Land Use Planning Act." It limits economic growth or permits it based on the relationship of emission sources on the land to the quality of the air above it. The ultimate impacts of clean air "land-use" legislation are still more obscure than the impacts of the creation of the interstate highway system on population shifts and factory and shopping center development. The recent sharp increases of gasoline prices and imposition of the 55-mile-per-hour speed limit may, however, ameliorate the savings of time and money made available by the interstate highways.

All these issues (interstate highways, land use planning, clean air, gasoline price escalation and the 55-mile-per-hour speed limit) have been more or less controversial because of direct economic or political reasons, not for the indirect impacts that have inexorably flowed from them.

Even with perfect foresight, there is room for a wide array of arguments on a controversial political issue, including the argument that the government should do nothing.

The nation's urban problems have been the result of forces which no government could stop or even substantially alter. The demographic trends, especially the movement of jobs and people to the South and Southwest, have had an economic impact far in excess of anything the government could have done to "compensate" cities for this out-migration. The problems caused by urban decay have required the resolution of too large and complex a group of complementary and contradictory interest for any neat or simple solution.

Not surprisingly, the one-dimensional solutions which have been tried have also failed to ease substantially the "limitless market" problems of our older cities. Those "easy answers" have included the Model Cities program, which was launched with great fanfare in 1966, only to be abandoned in favor of community development block grants in 1974. They have included massive welfare housing projects, which have caused social disintegration in many low-income neighborhoods. In one celebrated case, the Pruitt-Igoe project in St. Louis, a large low-income housing project simply had to be torn down after only 20 years of use. They also included the Department of Housing and Urban Development with its proliferation of ad hoc categorical assistance grants. In each of these cases, the major result seems to have been the creation of new jobs in Washington and the increase in time and expense by cities to keep up with federal red tape.

The other side of the coin, of course, is that this country is in the middle of social revolution as dramatic as the westward movement of the nineteenth century. Some federal policies, especially those that have facilitated long-distance travel or certain regional advantages, have hastened this revolution.

Whether one cites the New Deal efforts to modernize the South in the 1930s, the migration from South to North in the industrialization of the North in the decades before that, or the federal defense investment in the South during and following World War II, the momentum of change has moved far beyond any original stimulation. In spite of the impetus, the revolution and the social innovations it has engendered, such as nationwide motel chains, large retirement communities, and network television programming, have been largely a private sector evolution. In this complex policy area the government has been less successful at social invention than the private sector. The Social Security system may have made retirement more secure, but it does not alone explain Sun City, Arizona.

Federal Budgeting Policy: You Lose a Few, But Then Again You Win a Few

The methods that the federal government have developed to cope with budgeting technology are interesting as examples of how new technology, including new theories about the way government spending affects the economy, has brought on corresponding changes in political institutions.

Prior to the Budgeting and Accounting Act of 1921, the various executive branch departments submitted their own appropriations requests to the Congress through the Secretary of the Treasury, who was required to prepare a "book of estimates" of federal spending. This fragmented procedure (each department lobbied directly with Congress for its share) was clearly inappropriate for a country which had just experienced a large surge in war-time spending. The 1921 Budget Act required the President to submit the budget to Congress and created the Bureau of the Budget (now the Office of Management and Budget) to manage the preparation of the budget document. With certain changes in accounting procedures, this bureaucratic system remained essentially intact until recently. The Budget Bureau itself, however, expanded considerably from an appropriation of $5,000 in its first year to $10 million by 1969. This growth was related to the growth in the size and complexity of the federal government and the introduction of technological innovations such as computers and sophisticated econometric techniques.

By the end of the 1960s, however, Congress had become increasingly concerned about the uncontrollable growth in federal spending. In part, this growth had been brought about by the Vietnam War. But we subsequently learned that even with the winddown of military spending after the war, federal spending would continue to climb. The growth was caused to a greater extent by the proliferation of federal entitlement programs and other forms of federal nonmilitary spending. Defense spending for goods and services (excluding military pensions) in fact fell as a percent of total federal spending during the Vietnam years (1968 to 1974).

Congress made several attempts during the late 1960s and early 1970s to bring spending under control. Most of them involved legislated spending ceilings. Because the ceilings contained significant loopholes and were unenforceable, they were ineffective except as expressions of congressional concern. Or perhaps they were embarrassingly clear indications that, while Congress may have "viewed with alarm," it lacked the collective political character to take any adequate action to restrain its own profligacy.

The issue was complicated by procedural and political factors. On the procedural side, Congress appropriated funds through individual bills, each one decided without consideration of the others and with little consideration to federal spending as a whole. This contrasted with the president, who had the resources of the Budget Bureau to produce a comprehensive and internally consistent annual budget, but who was justifying a lack of political restraint with the theory that "full employment budgeting" would create full employment. The relative lack of organization of the Congress vis-a-vis the President was a factor in the legislation that was subsequently enacted.

On the political side, the spending programs themselves created constituencies in the Congress, the executive branch, and the private sector that were dedicated to program expansion and refinement. Because some groups would inevitably be disadvantaged in the event of any budget restraint, it was always possible to muster enough resistance against individual cuts to persuade a fragmented Congress to keep the programs at previous real funding levels by allowing increases to cover inflation and often to argue for real increases to cover expanding needs— some demonstrable, some imagined.

One way of describing this situation is that the budgeting process had moved beyond the capability of Congress and the executive branch to restrain it. The institutions did not exist to bring federal outlays down to desired levels. Despite the consensus on the need to do something, federal spending continued to climb as a percentage of GNP. Again, Congress was in the position of being reactive rather than pro-active, and it was not ready to react to growing danger signs in the general economy.

The Congressional Budget Act of 1974 was designed to try to rectify these problems. By creating a Congressional Budget Office, it gave Congress a budget analysis capability that had previously been enjoyed only by the President. The House and Senate Budget Committees established by the Act enabled Congress to take a look at aggregate spending and enact a budget that was consistent with an overall fiscal policy. The two committees also comprised a force within the Congress for policing the actions of the Appropriations Committees in the event spending legislation was reported out that was inconsistent with the aggregate numbers.

The Budget Act procedures were critical to the achievement of President Reagan's forcing Congress to alter the patterns of federal outlays. In the future, Congress is likely to give macro-fiscal policy (the overall effects of tax and spending measures) its appropriate weight in the consideration of individual revenue and appropriations bills.

Another Budget Concept

In an area of similar impact on the total economy, the rapid growth of social regulation in the past 15 years makes a strong case for a regulatory budget.

In the pre-regulation past the fiscal budget may have been adequate to show the impact of government on the economy, since almost all federal government activities involved direct taxation and direct spending. If one added to these the financial commitments (through loans, guarantees, and insurance) of some 14 "off-budget" agencies, one could get a fairly clear picture of the government's influence on the economy. But with the recent rapid growth of the new regulatory agencies—the Occupational Health and Safety Administration, the Environmental Protection Agency, the National Highway Traffic Safety Administration and many others—the fiscal budget no longer conveys a complete picture of the government's impact on the economy. Most of the economic effect of regulation is hidden, since government-required, private sector spending for auto safety, mine safety, pollution control, and consumer protection, plus the attendant paperwork costs, do not appear in the government's budget figures. They are cloaked in "off-off-budget" spending required of the private sector to comply with federal regulation.

A clear example of the need for a budget showing the economic impact of regulation on the society may be seen in the environmental regulation of electric utilities. The massive cost of a smokestack scrubber to achieve cleaner air is passed on directly to the consumers, who must pay higher utility bills as surely as they pay taxes. But the federal budget fails to show these regulation-caused higher prices. It also fails to show the higher prices consumers must pay because of economic regulation by such agencies as the Interstate Commerce Commission, the Civil Aeronautics Board, and the Federal Communications Commission. The costs and benefits of both social and economic regulations should be more clearly available to policymakers and the public.

If these costs were minor, their omission from the budget would not be a serious problem. But they are not minor. The cost of complying with federal regulations presently totals about $100 billion a year, a cost equivalent to the entire federal direct-spending budget when John F. Kennedy was President 20 years ago. And these regulatory costs are growing. It is important, therefore, that the Budget Act of 1974 be amended to require that Congress annually establish a regulatory budget, along with the fiscal budget, to set a limit on the costs of compliance each agency could impose on the private sector in any one year. The timetable and the process provided for developing a regulatory budget would be similar to those governing the fiscal budget.

A regulatory budget would provide an incentive for the regulatory agencies to limit the compliance costs imposed by their regulations. It would certainly make the Congress and its agencies more conscious of the regulation costs they impose on producers and consumers. A regulatory budget, added to the fiscal budget, would provide a more accurate picture of the federal government's total impact on the U.S. economy and the public sector burdens it carries into world competition with other national economies. This would allow Congress to determine how much of the nations's output is to be devoted to public uses and how much is to be devoted to private uses. It would make possible a better balance between regulatory programs and traditional government spending programs. It would enhance the protection of the public's health and safety by requiring that the federal government establish consistent and rational priorities in pursuing regulatory objectives.

Emerging Issues*

Because technology will continue to develop at a rapid pace, Federal policy must continue to consider the problems that technological advance creates. There is probably little Congress can do to eliminate the

---

* The following discussion is adapted from a rather lengthy list recently published by the Congressional Clearing House on the Future, a group founded in 1976 within the Congress whose purpose is to provide information about emerging issues.

political difficulties inherent in this exercise. As the following issues show, the same principles which have made, say, an urban policy so difficult to implement will raise their heads in other areas as well.

The issues fall into four major categories, plus a miscellaneous category. The first two (energy/food/natural resources and man-made disasters) contain issues that have been in the public domain for many years. The second two, however, (genetic engineering and the problems of the information/communication age) are peculiar to the late twentieth century. They indicate how new technologies can quickly change our lives and just as quickly raise new public policy questions on how to adjust social values.

Energy/Food/Natural Resources

The OPEC crude oil price increases dramatized a debate that has occupied inquiring minds for decades: the political conflict between dwindling resources on the one hand and growing world population on the other. The debate has both national and international implications, since the strategic posture of the United States is heavily influenced by the availability of dwindling supplies of energy and other resources from domestic production and friendly as well as hostile nations.

With regard to energy, the new technologies which will influence policy include synthetic fuel production, on-site coal gasification, clean combustion techniques for coal, exotic (or not so exotic) new technologies such as magnetohydrodynamics, bioconversion, solar satellite systems, geothermal, solar, alcohol fuels and wind power. Breakthroughs in energy conservation could come in the form of computerized home energy regulation systems and pipelines to carry some foods such as grain.

Through the allocation of research and development appropriations, Congress can influence which technologies will be given priority in the decades ahead. Debate is likely to center on the technologies that offer the fastest payoff per dollar expended and show promise of being commercially developed without the need for government assistance.

While certain energy sources may be interchangeable, the fact that others are not is a factor of great significance in the problem. Electric home heating can be a substitute for home heating from direct burning of oil, gas, coal, or wood. And any of those fuels plus nuclear power can produce electricity. But America's transportation system runs on oil, and to replace the gigantic infrastructure of transportation (or agricultural production) with electric or coal-fired energy would indeed be a revolutionary change. Similarly, the petrochemical industry cannot be converted from oil or gas. The wood or coal-fired locomotive could return without much social impact, but the electric automobile of limited range and speed or the return of the interurban would alter American life drastically.

Enough work has already been done by the Joint Economic Committee on the subject of R&D to indicate that with the appropriate incentives and government backing, U.S. energy needs can be met during the coming decade without drastic infrastructure modifications. Most studies assume, however, that nuclear plant construction will proceed on schedule in spite of the political opposition, that crude oil prices will remain high, although a major slump in prices could remove the incentive for alternative domestic energy development, and that military emergencies will not preempt our oil supply. For political and other reasons, these cannot be considered foregone conclusions.

Among the specific questions which members of Congress have explored are: Should Congress mandate national energy conservation standards for the construction of new buildings? Should Congress require the use of alternative energy technologies on military bases and in federal buildings? Should it aggressively pursue the buildup of a U.S. strategic petroleum reserve? What should be the mix of Federal and private energy R&D? How should Congress balance the conflicts between energy and environmental factors with regard to coal burning, shale oil development, and nuclear waste disposal? What organizational structure can administer national energy policy (e.g., Department of Energy, Synfuels Corporation)?

As the questions suggest, even public policy issues with a high technological content must be resolved eventually through the application of socially inventive political and economic judgment, sometimes in the marketplace and sometimes through public authority. The Congress seems to have learned that neither is infallible, but the marketplace and the adaptability of individuals to make social changes in response to economic changes are often quicker, more flexible, and imaginative than those in authority.

Food technology is also likely to receive considerable congressional attention in the decade ahead. New technologies which show promise of altering food production include aquaculture, new forms of pest control, food additives which use polymer leashing or genetic techniques to pass through the body without being absorbed, and genetic engineering for livestock. Each of these developments will raise questions of safety and the desired degree of federal involvement (e.g., aquaculture may be feasible in small-scale operations which raises the question of whether the federal government should encourage its development through direct subsidies or tax benefits). Aquaculture, incidentally, has been claimed by some to be a feasible large-scale protein source for

developing countries. This technology, therefore, has important implications for U.S. foreign assistance programs.

Food transportation is an issue affecting marketing costs and energy
consumption. Although food pipelines may not be feasible until the
relatively distant future, they could hold enough promise for energy
conservation and therefore warrant special attention from Congress.
The construction of such pipelines would raise the same kinds of policy
issues as those involved in oil pipelines (e.g., environment, anti-trust,
and economic effect on other regulated transportation). Obviously, the
mere existence of the technology does not guarantee that its development can proceed smoothly. If energy conservation becomes a more
urgent economic necessity and national priority than at present, both
private industry and Congress will be hunting for other ways to conserve through low-energy forms of transportation. History has found
an economic place for dirigibles and wind-assisted ocean-going ships.
The political questions revolve around using federal incentives to develop alternative systems, as America did with railroads; also whether
competitive relationships in the transportation industry should be substantially altered as the nation has with inland waterways, and whether
there are compelling environmental or safety issues (e.g., trucks versus
pleasure cars on the superhighways).

The crunch between growing needs and dwindling resources extends
beyond food and energy to include mineral resources as well. More
important, it includes water. This country could find itself faced with
severe regional water shortages as population builds up in areas that
are naturally arid or semiarid or where energy is developed by water-
consuming techniques. On the other hand, scientists claim that certain
weather control techniques could be developed within the next ten years,
and patent applications have already been filed on machines which
extract water from the air. The prospect of "stealing" rain from one
region to moisten another has not only national but international implications. Congress has had to legislate in the past with regard to the
diversion of water flowing in the nation's rivers; it may find itself having
to legislate with regard to water in the air as well.

Mineral and other raw material availability may be enhanced by mining the seabed. Here important international considerations will be
weighed against need and technical abilities, but many cite the environmental problems of seabed mining. Some seabed mining is already
taking place, and the Reagan administration has at this writing decided
to take a new look at the Law of the Sea Conference. The commercial
success of seabed mining technology will depend to a large extent on
the steps which the Conference and the Congress take during the next
several years.

Man-Made Disasters and Other Safety Issues

Modern man-made disasters conjure up images of nuclear accidents or nuclear waste disposal. These will continue to concern lawmakers during the 1980s. A list of disasters or other safety problems brought about by relatively recent technology also shows the possibility of varied congressional response. A short list of modern man-made disasters goes beyond the relatively benign issue of nuclear generation of electric power to include such inventions as jamming devices which could knock out international communications systems, new forms of technical and biological weaponry (including genetic weapons), low-level radiation from certain kinds of equipment, new evidence about the dangers of "second generation pollutants" from waste disposal problems, the alleged "greenhouse effect" of carbon dioxide caused by fossil fuel combustion, concerns about the alleged deterioration of the earth's ozone layer through the use of fluorocarbons, and the possibility of completely new ways of killing one another through lasers or "force fields." Some of these developments could require international agreements or federal laws controlling even more stringently some of the pollutants which can be proven to cause longer range pollution problems. One report states that Congress may have to set standards on the use of combustible polymers, which if burned in large amounts could substantially poison the air downwind from a building where these materials had been used.

Genetics

The two categories discussed above show that a congressional response may be required for some of the emerging technologies in the fields of energy, food, water and mineral supply, and man-made disasters. Although the technologies may be new, the problems are not. The nineteenth century economist David Malthus first popularized the concern that the world's population would someday outstrip the world's resources. While there are obvious imbalances in food patterns, world ability to produce food has stayed ahead of consumption. And in spite of the invention of gunpowder and finding new ways to eliminate one's fellow man, it appears the population continues to expand.

The rapidity with which genetic engineering techniques are developing and the new applications for their use are being found, indicate that Congress will either respond to this issue on many fronts and in new ways or watch dramatic changes in biological science occur without addressing them. There have already been debates about the ethics of certain kinds of genetic research. Congress may also have to consider the safety of certain genetic products and the rules by which government should regulate genetic material. A revolution of social, economic, and philosophic proportion appears to be underway, with or without government participation.

Genetics has applications and raises policy issues in health care, food additives, forestry, gasohol production, and patent and copyright law, a list which alone means that at least seven different current government agencies would be involved in regulation. This suggests that genetic problems could become as administratively cumbersome as the problems of energy policy in the early 1970s. As most analysts will attest, the Washington departmental organization of energy policy is one of the weakest links in that policy today.

The Information Revolution

Computers are not new in the American economy, and the legal, political and social issues which they raise are not entirely new either. As the miniaturization of computer hardware and innovations in computer software bring computers within the reach of the average person, the legal and socio-political issues become more controversial and complex. Computers represent a revolution of far-reaching proportions in the fields of information transfer, commerce, and banking. Already they have made possible a level of commercial activity that would have been impossible using pre-electronic accounting methods a few years ago.

Congress has already had to legislate to protect against invasions of privacy. The Federal Privacy Act, for example, entitles one to access and change the data stored by certain public and private organizations. According to some observers, Congress may also have to consider expanding the principles of the Privacy Act, to guard against the misuse of new "snoop" technologies for investigating one's credit rating or other aspects of private lifestyle.

Computers as applied to information gathering, storage, and retrieval have many applications which could be of direct concern to Congress. The Library of Congress is putting its card file on computers and phasing out the old paper file. Other possibilities in store for the Library and for the operations of Congress itself include electronic books, video versions of the *Congressional Record*, automated congressional offices to deal with constituent mail, an information system on the status of statutes and policy issues, and field hearings by teleconference. In short, computers are in the process of markedly changing the operations of government as well as business.

Banking is another field where legislation may be necessary because of the information revolution. The use of touchtone telephone billpaying and electronic funds transfer, together with operational changes in the banking industry such as NOW accounts, threaten to alter substantially the traditional competitive relationships among the various sectors of the banking industry—relationships that have been carefully guarded for years. In the process some sectors may gain what is seen by others as a competitive edge. Despite the fact that the Congress last year passed an important piece of banking law reform, there is likely to be new legislation during the next decade. Depending upon how forward-looking that legislation is, the banking industry may be very different by 1990 from what it is today, both in the kinds of banking institutions that exist and the kinds of services they offer.

Health, Crime, Work: You Name It, New Technology Affects It

According to reports issued by the Congressional Clearing House on the Future, each congressional committee will be coming up against issues where it must consider the social and political effects of new technologies in the areas under its jurisdiction. The following list, while not comprehensive, indicates the variety of problems caused by new technologies and the kinds of political response that may be debated.

Medical Services. New medical technology is expensive. Should access to this technology be available only to those who can pay for it, or should taxpayers be footing the bill for what could be extremely expensive medical care for the poor and elderly? Should Congress be more involved in health planning so that communities do not get overloaded with sophisticated health care equipment, or is there a "dead hand" holding back competitive progress when government engages in too much local health care planning?

Safety Technologies. Developments like the automobile air bag often create strong pressures on government to mandate the widespread manufacture and use of new safety products or restrict the sale of others for reasons of safety, economic cost, or competitive relationship. Auto bumper standards are one example. But what about electronic "stun guns" which could replace police revolvers? Should they be required, prohibited, or left to the marketplace?

Transportation. Transportation technology could take this country in the direction of "dial-a-vans," light rail systems, and moving sidewalks in city shopping districts. The use of these systems, especially the more expensive ones, will depend in large part on the willingness of Congress to vote large appropriations for their development.

Technology Applications.   The transfer of federally funded space or defense technology to civilian uses has been widely acknowledged as useful but raises important issues of how technology developed in federal laboratories should be brought into commercial markets.

Workplace of the Future.   Where robots take the place of assembly line workers and new communications techniques revolutionize the nature and productivity of work and workers in certain industries, how should economic and social impacts be dealt with? This could bring about changes in the workplace and also raise policy issues in the area of labor law and industrial relations. Imaginative social inventions will be of critical importance in resolving such issues.

Concluding Remarks and Observations

The "social imperatives" referred to in the title of this volume are an exclusive concept for the practicing politician. On one level, the concept seems clear in its scope and implications. It is generally agreed that government should encourage technological change in the interest of the common defense and general welfare. The idea that technological progress raises social or political issues which must be addressed by policymakers seems almost too obvious to dispute. The corollary, that government should attempt to anticipate new technologies so that it takes the necessary political steps to encourage social inventions, is more debatable and clearly more difficult. But it is an objective to which both Congress and the executive branch have devoted considerable efforts—sometimes even when there has been no imperative of technological change requiring it.

Political leaders should be asking whether the institutions by which we legislate are sensitive and responsive enough to be capable of linking synergistically social invention and technological innovation. Much of this essay has described barriers that the politician encounters. The identification of these barriers, however, also indicates where attention should be paid to improvements. We should attempt to ascertain why these barriers operate the way they do and suggest some solutions.

Perhaps the major impediment to legislation dealing with the need for social invention is the focus of politics on immediate short-range solutions. Generally speaking, political issues arise from the perceived needs of citizens rather than from the findings of people who analyze the past, present and future. This political reality was perceived by America's founding fathers when they established six-year terms for Senators to balance the two-year terms for Members of the House of Representatives and life appointments for Supreme Court Justices. Other more recently developed government institutions whose stability is important to our political-economic system have a similar protection of tenure. For example, members of the Board of Governors of the Federal Reserve System enjoy 14-year terms, and the Comptroller General of the Government Accounting Office enjoys a 17-year term.

The presidency itself is one aspect of the problem. This was recognized by the framers of the Twenty-second Amendment, whose purpose was in part to limit a president to one re-election campaign. Some have argued that the Twenty-second Amendment is not enough, that the presidency should be set at a single, six-year term in order to free the president from the political temptation of using the first-term trying to win re-election to the second term so he can be a statesman in the second term. (In fairness, others argue that the Twenty-second Amendment makes lame ducks of presidents in their second term, and it is also suggested that second-term presidents spend time trying to assure party succession.)

For technological innovation to be matched by political action to stimulate social invention, a large number of people must perceive the need, the cause-and-effect relationship, and the desirability of the social change being made. Ideas can gather steam for years before they are held by enough people to become the object of serious political debate. For example, the proponents of supply-side economics had been writing about the subject for several years before it was generally recognized as a valid guide for economic policy. The problems resulting from past patterns or inaction have to be seen as so bad as to be intolerable. Doubtless there were scientists whose experiments indicated some connection between cigarette smoking and cancer decades before Congress debated whether to place a warning on cigarette packages and advertising.

The maturation of ideas, from the time they appear in a scientist's notebook or an economist's essay until the time when Congress takes some action to implement them can be a long process. Those who write about the subject assert that political leaders enter this maturation process relatively late. Prior to an idea becoming a hot political issue, it generally has appeared in technical journals, specialized seminars, and even the national press. By the time an idea has hit the popular press, many lobbyists feel it is too late to try to stop it if it is perceived as a solution to a problem of sufficient concern.

It may be that the long maturation process is necessary for refining an idea to the point where it can be effective in curing a problem and achieving sufficient political support for its implementation. But institutions have been and can be devised to shorten the process. The Joint Economic Committee of Congress, the Congressional Budget Office, and the Office of Technology Assessment all can be looked upon as instruments for taking "young" ideas and bringing them to the attention of the political establishment. The Joint Economic Committee, for example, was instrumental in the 1975-80 period in refining the concepts, weaving the intellectual rationale, lending institutional respectability, and developing public awareness of supply-side economics. Earlier it also helped frame the procedures which resulted in the Congressional Budget Act.

In this context, the educational function of these institutions is what makes them important. They educate Members of Congress. Even if they cannot educate the public as a whole, they can be part of the marketing process of a worthy concept. However, it will take a considerable amount of marketing to bring the American body politic around to the kind of reasoning this book represents. Long-ingrained social beliefs and practices are not easily altered, no matter how sound the logic or punishing the economic consequences. Liquor prohibition in America and family planning in India are two cases in point. Those who try intellectually to anticipate events have fewer constraints on them than those who must enact legislation to cope with perceived future contingencies. For political leaders to be aware that they are accountable to the future and for them to respond to social needs is as difficult as knowing what those needs are likely to be. But beyond that, there is the difficult challenge of successfully marketing an unproven answer to an unperceived problem. A knowledge of technological innovation, together with good intentions, does not guarantee that legislation will reflect the important connection between social inventiveness and technological innovation or that it will be politically successful, even if it is the most efficient answer.

A good example is offered by the problem of computer-fed automation of the workplace. This issue had been featured in the presidential campaign of 1960; it was then called the "cybernetic revolution." Despite the voluminous literature on the subject, the cybernetic revolution was slow in coming. Whether it was the lack of economic pressure for change, resistant work roles and union contracts, lack of management innovation, or some other infrastructural block, many industries were slow to change. Printing and publishing "went offset" in a rush because of competitive cost pressures. But other industries, like auto manufacturers, are only beginning to retool today. The developments of robotic technology, automated instrumentation, word processors, and low-cost computer terminals have each had their impact. Yet the predictions of what will happen to work patterns in America are at best anecdotal and at worst contradictory. They provide no firm or sure basis on which to enact legislation designed to bring social institutions up to date with this technological revolution.

But thoughtful legislators, Government officials, union leaders, corporate officials, academicians, and others are thinking about these issues. A recent report of the Committee on Science and Technology of the House of Representatives, for example, details the implications of new technology for the quality of urban life, the workplace, privacy, mental health, and many other items of social importance. Like predictions about the economy, however, they are not an indisputable guide to the future, but rather a set of educated guesses about the kinds of things that might happen and should be pondered by educators, tax writers, and investors.

A "solution" to the problem of finding out what America's social needs are going to be goes beyond simply improving forecasts, although work needs to be done in that field. How one reacts to the forecasts is equally important. If one responds to the problem of the moment, without regard to the lessons of history, even the best forecasts might yield less than desirable results. With regard to automation of the workplace, for example, history tells us that irrational fear of automation can have significant and far-reaching results. The politician who ignores lessons such as these is taking huge risks.

The Luddites who destroyed the factory machines in the early days of the Industrial Revolution were right about one thing: The factories would eventually destroy the handwork of cottage industry. But they were dead wrong about the economic threat to the standard of living of the common man. Automated production has yielded more and better goods and services for all, though not without social and economic upheaval. Farmers who welcomed the improvements in agricultural technology of the last century were slow to perceive the inexorable impact on the social institution known as "the family farm." The size of farm families shrank along with the shrinking parity of farm/city living standards, but all living standards improved as farmhands became factory workers.

It is not wrong to worry today about the future employment of auto workers in a robotized world, unless it causes a huddle mentality. The worry can be positive if it addresses what must be done about retraining, company/union retirement programs, and community planning. Growing employment in service industries are already providing part of the answer to declining manufacturing jobs.

The change in public attitudes about factory employment raise "chick-

and-egg" questions about social and economic change that are very interesting. Did the attitude change come before or after "guest workers" were imported to take certain jobs while established residents preferred other positions? Has the declining birth rate of an affluent society—a society less enthusiastic about physical labor but more interested in physical recreation—been the cause or the effect of technological innovation which reduces employment requiring physical effort?

Politics dealing with only the present, and politicians ignoring evidence available today about the future, ill serve both the present and future constituencies.

But political issues brought about by technological change are often larger than the technological ones themselves, and that makes politics a tricky business. The political process in a democracy is the means by which people with opposite viewpoints and conflicting interest resolve their differences. That means that political leaders will sometimes disagree on recommended policy even when there is perfect foresight about where the future will push policy.

The issue often comes down to whether the government should be part of the solution, should attempt to legislate and regulate a solution to the issues, or leave the solution up to the private sector. Federal legislation and regulation do not have to be counterproductive, although all too often they are. We could budget regulatory costs to the society the same way we budget expenditures and taxes. We could try to prevent regulations from being counterproductive or contradictory. And we could require all regulations to be cost effective. President Reagan's regulatory reform program in fact is a conscious step in that direction.

If we assume that government regulatory incentives are a system of rewards and punishments, then the role of government in punishing success by high taxes and burdensome regulations, and the rewards by government given to inability or indolence through food stamps and welfare benefits, are well established. Whether they are desirable in the short and long run is a more difficult issue. But that question hinges on whether social and technological progress has been achieved to the desired degree on our past and present system of incentives.

In political life the issue can never be neatly phrased. This is because on any particular point of political difference it is almost always possible to find "experts" on both philosophic sides of the issue. This is because issues frequently turn on tradeoffs between economic and noneconomic values such as the environment. As a result constitutents will often respond to sensationalist news articles on one side or the other in such a way as to generate significant political reaction.

This honest difference and the ability to excite passions over the differences are both the weakness and the strength of politics in a democratic republic. For, in a sense, democracy means having the freedom to choose a course of action for or against one's long-term interest. The strength of this principle is that a nation's vitality is enhanced when its citizens realize they control their own destinies and act responsibly—whether that responsibility is seen as a long- or short-term interest. Like the ancient Greeks, we can and do attempt to look into the future. And our country is undoubtedly better at it, or at least more sophisticated, than the Delphic Oracle. The stability of our nation still depends upon the wisdom with which it responds to actual and anticipated events. Sometimes it is necessary to react strongly to soundly address a problem. But at other times prudence dictates patience to let nature take its course. And in other circumstances, our failure to anticipate may find us the victims of inexorable events. The exercise of this judgment is the essence of politics and the only fundamental truth is that each of the three courses will be appropriate at different times.

chapter eight

ENVIRONMENTAL REGULATIONS AND TECHNOLOGICAL INNOVATION*

Robert D. Hamrin

Technological Innovation in the United States Today

S peaking woefully of the decline of American technology has become fashionable. Some observers attribute this downward trend to the web of regulations that allegedly limits companies' resources and strangles their incentives to innovate. This chapter disputes that theory and discusses what government can do in the 1980s to ensure that the regulations not only have a minimal inhibiting effect but also work increasingly toward stimulating innovation.

A General Assessment

Those who argue that the United States is in technological decline are able to marshall an impressive array of supporting evidence and statistics, mostly related to research and development (R&D) expenditures.

---

*From *Environmental Quality and Economic Growth,* copyright 1981 by and available from The Council of State Planning Agencies, Hall of the States, 400 North Capitol Street, Washington, D.C. 20001.

As a percent of the GNP, R&D spending peaked in the middle 1960s and declined until 1978. R&D spending for basic and exploratory research, measured in price-adjusted dollars, has been on a plateau for ten years. Because private spending for R&D from 1970 to 1977 was lackluster, its percent of the GNP was lower in the United States than in Germany or Japan.

Such statistics are part of a broader and more fundamental phenomenon which Jerome Wiesner, the recent president of MIT, refers to as "technological maturity." The argument—that much of the innovative thrust of corporations seems to have dissipated as several vital postwar industries have reached a mature stage—has also been sounded by business leaders. Jacob Goldman, group vice president of Xerox Corporation, believes there has been a decline in innovation and a shift to simply improving the old, while Jerry Wasserman, a senior consultant with Arthur D. Little, contends that most so-called innovations build on existing technologies and simply extend the state of the art. Jay Forrester, an MIT professor who was instrumental in computer development in the 1950s, is forcefully succinct: "Our present technology is mature. Since 1960 there has not been a major radically new, commercially successful technological innovation comparable to aircraft, television, nylon, computers, or solid-state electronics."[1]

Historically, theory maintains that as an industry matures, it becomes more resistant to new technology. The financial and organizational commitments to old ways of doing things are simply too great to make major changes. This argument has been used to explain why there were virtually no important technological advances in automobiles after the introduction of the automatic transmission in the 1930s until the new electronics of the mid-1970s and why Europe and Japan have surpassed America in steel technology.

Statistics from *Science Indicators* 1976 bear out the "building on existing technologies" theme. From 1953 to 1973, the fraction of major innovations that could be called "radical breakthroughs" declined from 36 percent in 1953-59 to 16 percent in 1967-72.

Although the overall assessment is generally somber, there are some bright spots. Industry's total R&D expenditures increased 16 percent in 1977 and 10 percent in 1978, reaching $21.8 billion. The Administration is committed to real increases in basic R&D expenditures in what is a largely "hold-the-line" budget framework. Most significantly, in recent years corporate laboratories have been turning out a stream of major developments, primarily in the electronics field but also in energy and biology.

Indeed, in sharp contrast to the maturity of basic industries, the electronics, communications, and information fields are bringing about what is increasingly referred to as an information revolution. While much of industrial technology was relatively crude, requiring only a modest scientific or theoretical base, the information revolution is the product of the most advanced scientific knowledge, technology, and management and represents a great intellectual achievement.

In the search for new sources of energy, the federal government and more than 100 companies jointly funded (at a total cost in 1978 of $2.7 billion) a number of projects to convert coal into cleaner burning or more easily transportable fuels, such as high-octane gasoline. Recently scientists have made major breakthroughs in nuclear fusion by using lasers, rather than magnetic forces, to focus the enormous energy needed to fuse hydrogen atoms and create a nuclear reaction. Other scientists believe the most promising energy research may lie in the large-scale production of hydrogen as a replacement for gasoline, natural gas, and other fuels.

Genetics and microbiology are also fertile fields for innovation. There is general agreement that microbiotics research has far-reaching commercial potential. Gene splicing may eventually be used to create new varieties of crops, as well as new strains of bacteria, for such purposes as converting garbage to methane gas or concentrating valuable metals from low-grade ore. In addition to these corporate developments, industry is making some of its most important technological gains in the soft technology of systems engineering. Systems engineers, the generalists of industrial research, create technologies by merging developments from unrelated fields in ways that might not be apparent to specialists. Outstanding examples are the space program and Bell Laboratories' pioneering work in glass fiber cables.

Government's Effect on Innovation: Mixed Review

Governments in all developed countries work to promote and shape technological development because it is the most important contributor to a nation's economic growth. A 1977 Commerce Department study found that from 1929 to 1969, technological advances were responsible for 45 percent of America's economic growth.[2] It also revealed that from 1957 to 1973, technology-intensive industries' output, employment, and output per worker grew 45 percent, 83 percent, and 38 percent respectively, faster than that of other industries; prices, however, increased only half as fast.

Yet, the private market has many inherent limitations in coordinating all the effects of technological change. For example, it can do little to control such adverse effects of technological change as unemployment, pollution, and unsafe products. Since private firms also cannot capture all the benefits arising from technological innovation, they will tend to underinvest from a societal viewpoint. Moreover, the limited scale of most firms prohibits their undertaking large-scale or risky developments. In other cases, the public interest will require a government role in shaping new technological development that private firms may not pursue (for example, pollution control research or transportation facilities for the handicapped and elderly).

Many, if not most, government programs were not originally intended primarily to affect innovation, but embraced a wide variety of independent societal goals. For the most part, policy toward technological innovation remains only implicit in many political decisions. Although no major, across-the-board support for basic civilian technology exists, there are three key forms of intervention: research, procurement, and regulations.

There is little disagreement about the positive effect of the first two forms. Outstanding contributions to industrial innovations from government basic research include its work on aerodynamics, high-temperature liquid cooling, and improved aviation fuel. Government procurement also has great strategic leverage in stimulating innovations in product areas with commercial growth potential, a prime example being establishing aircraft designers as commercial innovators.

The effect of regulation on technological innovation, however, is a highly debated topic. Discussion of this issue is often clouded by emotions raised by the larger issue of the desirability of regulation. Thus, considerable care must be taken, through a comparison of costs and benefits, to separate the treatment of the effects of regulation on technology from the societal value of regulation.

Popular writings generally inveigh against the stultifying effects of regulation on technological innovation, particularly the relatively recent barrage of environmental, health, and safety (EHS) regulations. Actually, very little empirical evidence regarding the subject exists, and the few studies that do attempt to measure these effects reach conflicting conclusions.

Perhaps the most definite conclusion that can be drawn is that general, sweeping conclusions concerning the effects of regulation on technological change must be avoided. Regulation seems to have both hindered and stimulated innovation. Much depends on the timing and quality of a regulatory intervention. If the needed infrastructure, which includes trained people, is not in place or is created concurrently to meet the regulatory requirement, severe dislocations may result. Also, intense pressure for rapid change may force industry to patch up an existing technology rather than risk the failure of a radical innovation. On the other hand, when regulation is steady and gradual and firms have sufficient time to comply, effective and innovative technological solutions often appear.

We also can conclude that because EHS regulation affects the industrial process at different stages of product development and production, it will have a variety of effects both on its stated goals of improved health, safety, and environmental quality and on technological innovation. For instance, regulations requiring companies to demonstrate product safety or the efficacy of products before marketing may produce effects different from those requiring such safety or control after marketing. And the effects of regulations requiring the control of production (process) technology, effluent emission or waste control, or the safe transportation of hazardous material will also vary.

Effects of Environmental Regulations on Innovation

The complexity of the regulation-innovation relationship precludes the possibility of deriving a final number that summarizes the net impact. Part of the complexity lies in the distinction that must be drawn between the direct effects on innovation in compliance efforts and the longer-term, ancillary effects on the general process of business innovation. Either effect is likely to vary significantly, depending on the nature of the regulation and the regulated industry.

As will be shown, environmental regulation has stimulated and retarded both forms of innovation. Since such effects cannot be summed, it is impossible to assert that the regulation has exerted an overall net change in the rate of innovation. What is clear, however, is that the innovation process has been redirected toward new social purposes.

The degree to which firms positively respond to calls for technological innovation depends largely on the incentives built into the environmental laws and regulations. The United States has given insufficient attention to using regulatory legislation to encourage the production of compliance technologies to achieve regulatory goals. Certainly the regulations themselves do not effectively stimulate such new technologies or production processes. This contrasts sharply with the experience of several other countries where approaches to regulatory design often focus specifically on new technology. Following is a critique of those provisions in the laws and regulations that relate in some manner to innovation.

Law and Regulations Are Not Innovation Oriented

Although a few provisions of the Clean Air Act and the Clean Water Act provide incentives for new corporate pollution control strategies, they have stimulated few innovative compliance responses. Once polluting industries are in compliance, they have very little incentive to seek improved environmental performance.

The provisions most directly related to innovation are the innovation waiver permits of the 1977 Clean Air Act amendments. With regard to existing sources, Section 113(d)[4] states that the EPA Administrator can waive the requirements of state implementation plans (SIPs) for up to five years if the polluting source:

1. proposes a new means of emission limitation.
2. can adequately demonstrate this new technology by end of the period.
3. promises that the new technology will result in either:
   a. better performance.
   b. lower cost.
   c. lower non-air environment impact.
4. cannot install and test the new technology without the waiver.

Section 111(j) outlines a similar program for new sources. The proposed system:

1. must receive a public hearing.
2. must have the state government's approval.
3. must not have been proved.
4. must have "substantial likelihood" that it will achieve either:
   a. better performance.
   b. lower cost.
   c. lower non-air environmental impact.

The waiver cannot run longer than seven years from insuance and four years after the new source starts operation.

The primary benefit of these provisions could be to allow the long-term operation of pilot or large-scale test facilities. However, since the Acts require compliance at the end of a period that is much too short to allow significant depreciation of capital, these provisions do not provide any real incentive to build a commercial facility that has potential benefits but the risk of not achieving performance standards. Reflecting this problem and the lack of publicity on these provisions is the fact that as of November 1979, only one application out of 23 has been granted a waiver.

Congress viewed the new source performance standards in the 1977 amendments as "technology forcing" provisions that would stimulate innovation and environmentally improved production processes. The requirement that new sources achieve performance corresponding to the best demonstrated technology was originally intended to provide the incentive of an assured market to developers and vendors of improved controls. The incentive has proven to be remarkably weak.

The phrase "technology forcing" describes a regulatory provision that goes beyond a voluntary incentive to bring forth new technology. The CAA amendments use it in three contexts: nonattainment regions, nondeterioration regions, and state petitions. The "technology forcing" provision (Section 173) requires new sources locating in non-attainment areas to use "lowest achievable" emission rates. Section 165 applies a "best available" criterion to control technology for any emission source in a nondeterioration area. After Congress modified the bill, the distinction between "best available" and "lowest achievable" was essentially nonexistent. Finally, Section 111(g)[4] allows a state to petition EPA to upgrade new source standards, with the availability of new technology as an acceptable rationale.

A final innovation-related provision of the CAA amendments is the offset policy for nonattainment areas. It deals with the key question of how to initiate a process that will stimulate industrial innovation designed to improve environmental as well as economic performance.

The offset policy is one of a number of economic incentives in effect or under consideration by EPA. Bill Drayton, EPS's Assistant Administrator for Planning and Management, has emphasized that offsets and such other economic incentives as the bubble policy, reinforced by the banking of reductions and the use of deal-making brokers, are critical for increasing the rate of innovation in control technology. In a recent study of economic incentives, two of the three advantages cited were that they create a better climate for technological change, directly encouraging it in some cases, and they encourage innovation.[3]

The 1977 Clean Water Act (CWA) amendments demonstrate the determination of Congress to advance the use of innovative and alternative technologies in sewage treatment. The amendments state that in providing grants or subsidies, EPA will prefer those systems that "incorporate waste-water reclamation and energy recovery, as opposed to the conventional concept of treatment by means of biological or physical chemical unit processes and discharges into surface waters."

The amendments contain three interlocking financial preferences, in the form of direct capital subsidies rather than tax subsidies for communities that invest in new technology for reutilizing sewage in publicly owned waste treatment plants (POWT). The major one, Section

202(a)(2), increases the federal cost share from 75 percent of a POWT's construction cost to 85 percent if a community uses an innovative approach. This provision is a strong incentive because it represents a 40 percent capital cost reduction for the localities.[4] A further inducement is the lower relative operating costs of nondischarge approaches—costs that localities generally have to pay in full. Under Section 201(j), an innovative grant application may show estimated life cycle costs of up to 15 percent more than the conventional guidelines technology and still be approved. Finally, Section 202(a)(3) provides that if an innovative or alternative technology project fails, the government may give the community a 100 percent grant to fund its modification or replacement costs.

The CWA amendments relate to innovation in three other ways: they require that innovative and alternative technologies be studied and evaluated for all projects, they provide a 15 percent credit for cost-effectiveness analysis in evaluating such projects, and they allow states to modify their priority lists to give greater preference to such projects.

Though the Toxic Substances Control Act does not have any innovation-specific provisions, Section 2(b)(3) contains perhaps the clearest legislative expression of the importance of technological innovation to those involved with both industrial production and improving worker and consumer safety or the environment.

> It is the policy of the United States that . . . authority over chemical substances and mixtures should be exercised in such a manner as not to impede unduly or create unnecessary economic barriers to *technological innovation* while fulfilling the primary purpose of this Act to assure that such innovation and commerce in such chemical substances and mixtures do not present an *unreasonable risk* of injury to health or the environment. (emphasis added)

The Act does contain a provision allowing companies to share testing costs for particular chemical products that may be an important benefit to small firms' innovation in the chemical industry by replacing much of the present uncertainty with preestablished and well-defined procedures.

Besides enacting the environmental laws, Congress has enacted a number of tax incentives for pollution control. On the whole, these tend to bias corporate decisions toward end-of-pipe conventional control technologies rather than innovative, preventive process changes.

The Revenue Code contains two major tax benefits that reduce the cost to firms of complying with pollution control regulations: (1) rapid amortization and the investment tax credit for pollution control hardware and (2) tax exemptions for municipal bond financing of pollution abatement facilities.

The Code's accelerated depreciation provision (Section 169) allows pollution control hardware to be amortized faster than other equipment. Although the deduction applies to both pollution abatement and prevention, the nonsignificant change requirements severely restrict the latter—namely, preventive facilities *cannot* lead to a significant increase in output, capacity, or the useful lives of equipment, a significant reduction in operating costs, or a significant alteration in the nature of the manufacturing process or facility. "Significant" is defined to be more than a 5 percent change. A related provision allows the investment tax credit to be claimed for one-half the costs qualifying under Section 169.

Often, firms do not use these provisions because they find that using only the full investment tax credit is more beneficial. When they are used, they tend to bias environmental compliance. Since companies can clearly identify only end-of-pipe technologies as pollution control investments (because they find it difficult to identify operations and maintenance expenditures that are included in process change control efforts as an investment *per se*), they tend to adopt those techniques that are easily subsidized (and thus made artificially less expensive) rather than more efficient change-of-process controls. Furthermore, the nonsignificant change requirement provides substantial bias toward incremental or patch-on rather than major process changes.

Two recent studies for the Office of Technology Assessment and the National Bureau of Standards support these conclusions:

> the current tax treatment for pollution-control expenditures appears to have simply encouraged retrofits on existing facilities instead of investment in newer and more efficient technologies.[5]

> the liberal tax treatment for investment in general appears likely to work against innovation in the short-run, having a greater relative impact on the expected return to conventional technology.[6]

Finally, EPA influences the utilization of control technologies through its research and development and its technical assistance. In the past, R&D concentrated on end-of-pipe control technologies that may have contributed to the underrepresentation of process change. Recently, the focus has shifted to process change, but it is too early to assess the impact. In the past, EPA's technical assistance efforts have also favored end-of-pipe controls. Since process changes are more case specific than end-of-pipe controls, EPA engineers asked by industry to provide technical assistance often recommend controls based on their general applicability.

Environmental Compliance Innovation

Though it is clear that regulations encourage technological change in compliance responses, such changes will not necessarily be "innovative." Indeed, the question of whether the technology-based approach has encouraged innovation is still an issue subject to much controversy.

Critics of this approach have argued that perverse incentives for technological innovation are inherent in the way technological requirements are framed. Company officials, for instance, may reason that if they do not meet the effluent limitations, they will be safe from EPA prosecution so long as they have made a "good faith" effort to achieve the standards by adopting the suggested strategy. If true, then the technology-based approach may encourage a risk averting strategy of adopting the suggested or sanctioned technologies, even when they may be expected not to meet the standards. Also, in cases where the regulatory standard is set at the level of the best practice in a few leading firms, the major effect that occurs is diffusion of an existing technology to the lagging firms rather than innovation.

On the positive side, it can be argued that by requiring dischargers to control their emissions to levels that engineers say can be achieved, a technology-based policy forces the adoption of the best available technology (BAT). Other counter-arguments suggest that: (1) sufficiently strong commercial incentives exist to produce high-standard equipment; (2) there is evidence that applying suggested technologies does not hinder innovation; (3) no firm is permanently exempt from BAT regulations; and (4) a non-compliance fee can be levied.

A recent study by the Organization for Economic Cooperation and Development (OECD) summarizes the present situation well:

> although technology-influenced effluent standards may appear to be determinant and objective, in practice, many controversial judgments and decisions appear to have been made in translating the legislative directive to base effluent limitations on technological considerations into specific numerical limitations. . . . The technology approach is *not* neat and simple compared to the messy complex value judgments that are associated with ambient environmental standards, setting effluent charges and so forth.[7]

Ancillary Innovation

As for the longer term, ancillary effects on the general process of innovation, the accumulating evidence suggests that environmental regulations provide a very healthy stimulus that is likely to outweigh any negative effects.

Looking first at the negative aspects, regulation can diminish conventional innovation by reducing the overall R&D budget by diverting R&D funds from conventional to compliance research and by changing the allocation of funds in conventional R&D activities. Although these effects are widely perceived to be significant, evidence on the magnitude of these effects does not exist. Furthermore, only a small fraction of the money spent on industrial R&D is going into pollution control R&D—3 percent in 1977.

Another alleged negative impact is that environmental regulations accelerate the technical and economic obsolescence of equipment, since the old equipment not only is generally more polluting but also necessitates larger expenditures to make it conform to the new standards. Though this effect does exist, it is not necessarily negative, since the regulations simply accelerate inevitable trends.

On the other hand, an MIT study confirmed that regulations can have a positive impact on innovation.[8] The study analyzed the regulation/innovation relationship in five industries (auto, chemical, computer, consumer electronics and textile) in five countries (France, Germany, Holland, the United Kingdom, and Japan). The study found that innovations for ordinary business purposes were much more likely to be commercially successful when environmental/safety regulations were present as an element in the planning process than when they were absent. In addition, compliance-related technological changes often led to product improvements far beyond the scope of the compliance effort. The study also concluded that many direct attempts of governments to encourage innovation were not correlated with project success to as great an extent as when regulation was used. Another study, by the Science Policy Research Unit of the University of Sussex, suggested that environmental regulation in the six countries studied does not appear to have prevented or delayed innovation to a significant extent and may even have had a "substantial stimulating influence on improving quality and performance." Finally, the OECD reports that environmental regulation has (1) helped Japan develop cheap alternatives to PCBs and devise new techniques to meet strict auto emission standards; (2) stimulated in Norway the development of a cheaper energy-saving alternative to open-furnace burning; and (3) stimulated an effort in France to recover many unrecovered pollutants for their economic value.

Looking more specifically at the United States, there is an often overlooked but very significant stimulus to general business innovation—process change—which often yields economic or energy-related benefits. The thesis is that environmental regulation often shocks firms out of a rather inflexible production system, thereby providing the catalyst for innovation to occur. A number of recent studies provide evidence supporting this thesis. The Denver Research Institute found "strong support" for the idea that the need to change occasioned by regulation provides an opportunity to make process improvements in areas unrelated to regulation.[10] A study of the effect of environmental regulations on the industrial chemical industry reported that 33 percent of its respondents cited indirect benefits relative to process improvements.[11] Another study of this industry found that the people interviewed thought that the new analytical capability needed to assess the health/ environmental risks of both new and existing products would be important for the future development of products and processes.[12]

It should also be recognized that the emergence of a new sector—the pollution control industry—shows governmentally stimulated technological innovation on a large scale. The industry's recent history (1972–76) reveals better than average growth, yet only average profitability. In 1977, the markets for pollution control products totaled $1.8 billion. These markets are projected to grow to $3.5 billion by 1983, a pace of 11–12 percent per year. Furthermore, the industry and the technologies it uses show a very favorable balance of payments, with the United States maintaining a position of great strength in world markets.

Regulation-Induced Process Change

While on the whole it may be concluded that most firms have not responded to regulation by implementing process changes, it is also true that environmental regulations have been the most important single force in the post World War II era causing American industry to rethink and change established production and management processes on such a wide scale. Moreover, while a large majority of firms have responded to the Clean Air and Water Acts' requirements by applying end-of-pipe controls (either out of necessity or unwillingness to engage in more creative thinking), a small but significant number of firms have evaluated compliance alternatives and have often made minor modifications in their production processes. In a growing number of cases, firms have thoroughly reevaluated their entire production system and have developed superior processes or process controls that not only solved pollution problems but led directly to the adoption of energy-saving or more economic processes. Based on a 1977 survey of industry, the Census Bureau estimated the value of the energy and materials recovered as a result of pollution control measures to be more than $950 million.

Most of the process changes are specific to a given company, but the auto industry's use of electronic microprocessors to control engine performance for both fuel economy and lower emissions levels provides a striking example of an industry-wide change. Philip Caldwell, vice chairman and president of Ford Motor Company, actually refers to the "opportunities" provided by government regulation that have initiated "revolutionary change" in auto design that are creating a "great new market."[13] He maintains that typical cars of the 1980s will be 100 percent more fuel-efficient than those of 1974, resulting from an "explosion" of new applications in electronics and other "explosions in materials substitution and the use of computer technology in design and manufacture."

The auto industry's adoption of microprocessors illustrates three important principles. First, the impetus that regulation can create for transferring technology from one industry to another. For example, a large market has been created for firms producing such devices as microprocessors and sensors. Second, new technologies sometimes provide great potential for expanding and improving their application, in this case for other electronic automotive applications such as controlled braking. The third, perhaps most important, principle is that actions that complement normal competitive pressures for change on an industry often appear to be more effective in promoting innovation than those that do not relate to market forces. Industry thus finds itself doubly motivated to innovate.

Following are examples of companies that have reevaluated their production processes, leading to the adoption of innovations that resulted in significant economic and or energy benefits:

Economic and Energy Savings

1. *Glass Containers Corporation* has developed the largest glass recycling program in the United States and has made technological discoveries that will increase recycling nationwide. By using a much higher percentage of used glass in a batch of molten glass and developing a computerized control system for using waste glass in the batch, the company reports that air quality improvement has been so great that no special pollution control devices have been necessary for furnaces, and that fuel savings and longer furnace life have resulted. This system provides an additional economic benefit to cities and towns faced with expensive landfill problems.

2. *Uniroyal Chemical,* to better dispose of hundreds of thousands of gallons of nonenes, devised a process to combine the nonenes with fuel oil and burn them in the company's steam-generating boilers. During the first year of operation, the process recovered and burned 366,000 gallons of waste nonenes, resulting in substantial fuel oil savings. Against investment costs of $48,000, first-year savings from the changeover were about $183,000.

3. *Long Island Lighting Company* used a magnesium fuel additive to reduce sulfur trioxide concentrations. Burning magnesium oxide with its Venezuelan fuel oil not only solved the environmental problem but also produced a marketable byproduct, vanadium. In 1978 the company sold 362 tons of recovered vanadium for $1.2 million and saved $2 million in fuel because of thermal efficiencies and $400,000 because of reduced boiler corrosion.

Economic Savings

1. *Gould, Inc.*, began replacing more traditional waste-water treatment methods with a reverse osmosis cleansing process that permits its plants to recycle most of the water they use. Savings from the system, which allows recovery of large quantities of copper, conservation of sulfuric acid used in plating, and achievement of several other economies, are expected to exceed $450,000 a year, thus recovering in less than two years the $845,000 system installation costs.

2. *Great Lakes Paper Company* installed an $8 million closed-cycle water treatment system, which is expected to save $4 million each year in lower costs for chemicals, water, and energy, while containing contaminated effluents.

3. *Hercules Power* spent $750,000 to reduce solids discharged into the Mississippi River. As early as 1971 it was saving $250,000 a year in material and water costs.

4. Process improvements in a *Gold Kist* poultry plant cut water use by 32 percent, reduced wastes by 66 percent, and produced a net annual saving of $2.33 for every dollar expended.

5. *Centron Corporation* developed a system to recover solvent vapors from the production of magnetic tape that were previously expelled into the atmosphere. Estimated savings are $50,000 a year, but actual savings are much higher since the firm can now use less expensive, nonexempt solvents in the process. (Air pollution from solvents has dropped from 450 to 59 tons per year.)

6. *J. R. Simplot* virtually ended its discharge altogether with a system combining primary treatment of its liquid effluents and spray irrigation. The system has eliminated from nearby streams and groundwater almost all the 40,000 pounds of biological oxygen demand, the nutrients, and suspended solids. Moreover, the plant now sprays the nutrients in the wastewater on the land to produce high protein forage, which it combines with other solid wastes from the plant to feed 26,000 yearling steers. The waste heat in the effluent used to irrigate the land allows a 10- and 11-month growing season and an annual yield nearly twice that of normal crop lands in the area.

7. *Dow Corning* found that a $2.7 million capital investment in equipment to recover chlorine and hydrogen previously lost to the atmosphere reduced operating costs by $900,000 a year—a 33 percent annual return on investment.

Energy Savings

1. *The Aluminum Company of America* developed a way to recycle fluoride in its refining and smelting process and thereby eliminated the need for huge quantities of water. In addition to significantly reducing fluoride fume emissions, Alcoa's new fluidized bed technology cuts energy consumption in two processes by 30 percent.

2. *Corning GlassWorks* devised an improved, electrically powered melting process that eliminated the loss of certain raw materials to the air. The process has vastly reduced air emissions and simultaneously allowed better control over glass production. Energy consumption has been reduced by a factor of 3.7 in the new plants, and the reduced space required for building this system has lowered construction costs.

3. *The Georgia-Pacific Corporation* developed a special scrubbing system to eliminate "blue haze" emissions caused by plywood production. Collecting the airborne pitch produced a thick liquid, which has a Btu rating that is the equivalent of #6 fuel oil. The firm now uses this residue as a fuel supplement, and the four units in operation collect enough residue to replace 51,000 gallons of #6 fuel oil each year.

4. *The Florida Power Corporation* installed burners that were designed to operate at very low levels of excess air. This development simultaneously increased boiler efficiency and lowered fuel consumption. It also reduced visible emissions from 40 percent to 10 percent (state standard is 20 percent). After one year, Florida Power's ten burners have consumed 4,000 fewer barrels of oil than with the previous system. And the new system reduces operating costs because it requires less manpower.

5. *Wheelabrator-Frye* found that converting garbage and refuse to energy could be a practical enterprise. The firm constructed a facility on a former land-fill site. The facility receives all the garbage from Boston's thickly settled North Shore and processes it, producing economical steam energy for a nearby General Electric plant. This system has made cheaper steam energy available and reduced costs to the North Shore communities in refuse disposal. In addition, it saves 27 million gallons of fuel oil a year.

The fundamental principle that all these examples illustrate so well is the strong link between management commitment to compliance and the nature of the effect of regulation. Strong management commitment normally results in both lower compliance costs and favorable effects on innovation.

One company that has had such commitment and results is the 3M Company. In 1975 it introduced the Pollution Prevention Pays (3P) program to stress conservation-oriented technology that will prevent pollution at the source in product and manufacturing processes rather than remove pollution after it is created. This is a continuing effort that focuses on eliminating pollution sources through product reformulation, process modification, equipment redesign, and the recovery of waste materials for reuse.

To date, 39 projects selected for recognition have annually eliminated the equivalent of 75,000 tons of air and 1,325 tons of water pollutants, 500 million gallons of polluted wastewater, and 2,900 tons of sludge. In three years, the total estimated saving is about $17.4 million, mostly from eliminated, reduced, or delayed capital pollution control investment and operating costs, including savings from improved products and processes and some sales retained from products purged of a pollutant or toxic compound. Ten overseas subsidiaries have their own programs that have undertaken 75 projects, representing an additional saving of $3.5 million.

An example of a 3P project involved recycling cooling water that previously had been collected for disposal with wastewater. Reusing the cooling water allowed the capacity of a planned wastewater treatment facility to be scaled down from 2,100 to 1,000 gallons per minute. The recycling facility cost $480,000, but 3M saved $800,000 alone on the construction cost of the wastewater treatment plant.

After three years' experience, the company is pleased with the results and is as firmly committed to the program as it was when it started. According to Joseph Ling, vice president for environmental engineering and pollution control:

> Necessity, the cliche goes, is the mother of invention. This may be an overworked phrase, but that makes it no less true—particularly in reference to pollution abatement. 3M's attention was directed toward pollution control initially because of necessity. Now it is directed toward elimination of pollution sources for the same valid reason—it is necessary. The cost of relying entirely on pollution removal technology—the little black box at the end of the pipe—simply is becoming too great, considering the myriad other demands on our financial resources.[14]

This is precisely the type of resource-conserving philosophy leading to action that must become widespread throughout corporate America in the 1980s.

Shaping an Innovation-Inducing Program

Encouraging Resource-Conserving Technology

The 1980s will witness a new approach to pollution abatement—developing resource-conserving technology that will both conserve resources and lower the cost of pollution abatement by working to eliminate the sources of pollution in processes and products before waste is created. If waste is broadly viewed as the "irrational use of more than necessary resources to fulfill human needs and objectives," then resource conservation becomes "a positive image for the application of technology to fulfill human needs and aspirations without destructive impact on the environment."[15]

Over the long run, this type of technology will be much more environmentally effective and cost-efficient than the traditional end-of-pipe control measures so heavily relied on in the 1970s. The main problem with the latter "removal" technology is that no matter how effective it may be, it only contains the problem temporarily. Such measures that are subservient to the immutable Law of Conservation—we can change the form of matter but it will not disappear—do not solve the problem but merely shift it from one form of pollution to another.

Removal technology presents a number of other problems. Control measures create "off-site" pollution, which is pollution generated by those who supply the materials and energy consumed in the pollution-removal process itself. Also, the cost of pollution control, the resources consumed, and the residue produced increase exponentially as removal percentages rise to the last few points. At this stage, eliminating the final stage of pollution often can create problems many times greater than those that were eliminated.

Thus, using only the control approach to solve environmental problems is contrary to the objective of a resource-conserving society, which should instead change pollution-abatement philosophy and adopt resource conservation-oriented abatement technology. A simple concept, but with profound implication, this change is no more nor less than the practical application of knowledge, methods, and means to provide the most rational use of resources to improve the environment.

Resource conservation technology means eliminating the causes of pollution before spending money and resources to clean up afterward, as well as learning to create valuable resources from pollution. In short, the concept is to use a minimum of resources and create a minimum of pollution.

There are at least four major environmental and economic benefits of an approach that more and more substitutes pollution-source elimination for pollution-control technology:

1. It allows us to save energy and other valuable resources that can be applied to other problems and provide opportunities for human betterment.
2. Since little, or in certain instances, no waste is generated, pollution problems are eliminated once and for all.
3. The new technological spin-offs could offer innovative opportunities to convert waste materials called pollution today into valuable resources for other constructive uses tomorrow.
4. It is the most effective, long-range solution to the increasingly serious and complex global environmental problems.

A resource-conservation approach also helps eliminate a type of pollution left untouched by the control approach: the environmental impact of products after they leave the factory. Since a product is likely to contain any number of pollutants, a problem is created for the user which is beyond solving by controls in the manufacturer's factory. If the user is another manufacturer, the "within the product" pollution could become a "within the manufacturing process" pollution problem. The vicious circle can be broken only by eliminating the pollutant from the product in the first place through resource conserving technology. For example, by removing a mercury catalyst from an electrical insulating resin, the 3M Company did away with any pollution problem the mercury created for the user. The new formula was more environmentally acceptable and helped 3M prevent a substantial loss in sales.

Resource-conservation technology is neither a panacea nor a substitute for pollution control. Some industries cannot easily change their processes without disrupting or halting production. Changeover may be too costly, or there may be no resource conservation technology to eliminate the pollution sources.

The goal should be to use resource conservation technology where and when it is possible and practical. Individual industries must apply their ingenuity to develop their own techniques relating to their special pollution problems. Over the long run, the resource-conservation approach should prove to be more environmentally effective and less costly than conventional control methods.

Implementing Legislative and Regulatory Reform

To encourage the development of both the new technologies necessary to achieve environmental goals and safer products and materials. Congress should resolve to increasingly consider issues concerning regulatory system design and implementation. The debate concerning environmental regulation, focusing on the need for new legislation and the *stringency* of regulatory requirements, has underemphasized questions relating to implementing the legislative mandates. Various ways of promoting the growth of innovative compliance technologies include government support for efforts to achieve regulatory goals through technological change, effluent taxes, and provision for joint research and development for pollution control.

Perhaps the most direct and effective action can be taken in drafting regulations. In the 1980s, there should be two corresponding policy goals vis-a-vis innovation toward which environmental regulations should work:

1. To the extent consistent with the environmental-safety goals established by legislation, regulations should not unduly hamper innovation for ordinary business purposes.

2. To the extent possible, regulations should encourage innovation in compliance and abatement technology. This innovation should both address alleviating the immediate hazard and contribute to longer-term systemic changes that can result in more environmentally benign and safer products and processes in the future.[16]

The goal is to design regulations that will advance both goals. Where these two goals are inherently incompatible, however, difficult policy tradeoffs will have to be made between them.

One of industry's most frequent complaints is that the compliance period specified in regulations is too short to allow for development of the most effective or innovative response. For example, the 3M Company states that "legislation and regulations usually provide neither the time nor the flexibility a company needs to develop products and processes which eliminate or reduce pollution at the source; the rules are directed primarily toward achieving pollution removal."[17]

The validity of this type of complaint is open to some question in that a several-year period of government scrutiny, hearings, and public controversy precedes most regulations. During this period, industry receives a quite clear signal of the standard to come. Nevertheless, it is true that developing important compliance innovations (especially those involving relatively major changes in existing technology) can take a number of years. In such cases, extending the time period between standard promulgation and full compliance (i.e., timing) could serve as a means of encouraging innovation. A closely related idea is to use more effectively use the experimental waiver provisions in the Clean Air and Water Acts. High-risk but strong candidates for advancing state-of-the-art technology should be more easily granted waivers in situations with low exposure and low adverse environmental hazard potential. Possible abuses can be controlled by specifying a limit on the number of waivers per industrial category or control type.

The experience of Glass Containers Corporation provides a recent example of the benefits that can accrue from flexibility in extending a compliance deadline. EPA first gave the company until June 1976 to comply with air quality standards. But when the company did not meet that deadline because it was having trouble finding enough used glass to use in its new recycling process, the EPA regional office granted it a six-month extension. Company officials enthusiastically responded to this flexibility:

> They could have come in here at any time and just said "do it now," and we might have cut our operation in half. Instead, they were patient, worked with us . . . now we are working full shifts around the clock, paying out well over $1 million to the local economy, keeping thousands of tons of glass out of the solid waste landfills, and turning out bottles as good as they have ever been.[18]

Much could be done in R&D to promote innovative responses and resource-conserving technology. Regarding federal R&D generally, any significant federal program designed to foster the development of new technologies should be required to evaluate the comparative environmental and economic advantages of the proposed technology. Regarding federal environmental R&D, the federal effort should be expanded to include work on improved generic methods for waste reduction, separation, and disposal. In addition, some industries sorely lack research and innovation on systematic approaches to simultaneous process and pollution control design to minimize overall production costs. Such approaches need to be developed for industrial processes, integrated manufacturing plants, and area-wide waste management.

Another broad government-wide reform concerns the contract and grant specifications of federal agencies. All federal agencies, particularly those like DOD, GSA, HUD, and DOT that purchase directly or control the purchase of manufactured goods can provide industry a major incentive to use processes that pollute less and develop products with the least possible environmental impact. This objective could be furthered by inserting clauses in the contract and grant specifications that favor government or government-subsidized purchases of the most environmentally safe products or of products derived from the most environmentally benign processes.

Another possible regulatory reform is to substitute for the substance-by-substance approach a "generic" standard for a class of substances. The generic approach greatly reduces regulatory uncertainty. Because it gives firms a clear and advance signal of the kinds of substances that will likely be regulated and the nature of the controls that will probably be required, it allows the private sector to develop appropriate, long-term technological options.

EPA should explore the merits of a privately administered revolving fund partly to insure against developmental cost losses when products are not registered (pesticides) or cannot be produced due to general regulatory provisions (chemicals). This would help alleviate the problem of companies that tend to hold back on developing new products because of their perception of the difficulty of getting new products registered or approved and the financial loss involved if their registration application or pre-market notification is rejected.

A recent report to the Congressional Office of Technology Assessment offered a list of regulatory alternatives that may foster innovation:

1. Expansion of direct government support for in-firm technological development in crucial areas (e.g., pollution control in automobiles) leading to both process and product change.
2. Modification of pollution control tax incentives, i.e., accelerated depreciation and municipal bond financing, so as to favor process redesign and the development of new products and materials rather than add-on modifications associated with purchasing of pollution abatement equipment.
3. Government financial support for major new technological advances when firms are unlikely to undertake them on their own either because such development would require large-scale efforts, would be long in coming to fruition, or their results nonappropriable (e.g., closed systems to contain toxic chemicals). This occurs in Germany and France as part of broader programs to encourage the development of new technologies for various social purposes.
4. Greater industry-specificity in standard setting (e.g., in the OSHA context) so as to minimize hardship when new technologies would be difficult to develop and to maximize health safety protection when the technological capacity is great.
5. Alternatives or supplements to standard setting , such as products liability or strict liability imposed on polluters, as in Japan.
6. A formal antitrust exemption procedure to clarify the status of joint R&D relating to environmental control technology.
7. Special programs to assist small firms' compliance efforts.
8. Effluent taxes as a means of achieving water pollution abatement on a regional basis (these have apparently been successful in Europe, especially in Germany, and are alleged to provide continuing incentives for more efficient control technology).[19]

What the preceding suggestions primarily illustrate is the need for a thorough reassessment of the means of achieving regulatory goals via technological innovation. Such a reassessment must be socially inventive and begin now if the United States is to become a resource-conserving society in the 1980s.

Notes

1. Jay Forrester, "Changing Economic Patterns," *Technology Review*, August-September 1978, pp. 52–53.
2. Cited in "Vanishing Innovation," *Business Week*, July 3, 1978, p. 47.
3. James Booth and Zena Cook, "An Exploration of Regulatory Incentives for Innovation: Six Case Studies," Washington, D.C.: Public Interest Economics Center, January 18, 1979.
4. It should be noted that since wastewater treatment funding has been without ICR systems for much of the grant program, firms have been subsidized for the dumping of their wastes. Thus, the 75 percent subsidy likely has led ultimately to a disproportionate use of end-of-pipe controls.
5. Center for Policy Alternatives, MIT, *Government Involvement in the Innovation Process* (Washington, D.C.: Government Printing Office, 1978).
6. Booth and Cook, *An Exploration of Regulatory Incentives*.
7. OECD, *The Influence of Technology in Determining Emission and Effluent Standards*, Paris, 1979.
8. Center for Policy Alternatives, MIT, *National Support for Science and Technology: An Examination of Foreign Experience*, CPA Document 75–12.
9. Science Policy Research Unit, University of Sussex, *The Current International Economic Climate and Policies for Technical Innovation*, November 1977, p. 20ff.
10. Boucher et. al., *Federal Incentives for Innovation*, Denver Research Institute, University of Denver, January 1976.
11. J. C. Iverstine, *The Impact of Environmental Protection Regulations on Research and Development in the Industrial Chemical Industry*, National Science Foundation, May 1978.
12. Center for Policy Alternatives, MIT, *Environmental/Safety Regulation and Technological Change in the U.S. Chemical Industry*, 1979.
13. "Caldwell Lauds Regulation as a Spur to Innovation," *Automotive News*, January 22, 1979.
14. Joseph T. Ling, "Pollution Prevention Pays," *Pollution Engineering*, May 1977, p. 33.
15. These definitions are from a seminar sponsored by the U.S. Commission for Europe on "Principles and Creation of Non-Waste Technology and Production."
16. These two principles were put forth by Ashford et. al., in *The Implications of Health, Safety and Environmental Regulations for Technological Change*, Center for Policy Alternatives, MIT, January 15, 1979, pp. 3–2 & 3.
17. Ling, *Pollution Prevention Pays*, p. 34.
18. Larry Kramer "Bottle Maker Cuts Costs, Pollution with Old Glass," *Washington Post*, July 2, 1978.
19. Center for Policy Alternatives, MIT, *Government Involvement in the Innovation Process*, pp. 59–60.

chapter nine

THE NATIONAL AERONAUTICS AND SPACE ADMINISTRATION: ITS SOCIAL GENESIS, DEVELOPMENT AND IMPACT

Karl G. Harr, Jr.
and Virginia C. Lopez

It could have been said, in the early 1950s, that the United States was ready for the space age—if one pulled together the existing bits and pieces of programs, mostly military enterprises, that characterized the effort up to that time. Thanks to large liquid-fuel rockets developed as military missiles, space flight was indeed possible. But to have said the U.S. was ready would have been stretching things.

As the fifties got underway, the United States was definitely behind the Soviet Union in emphasizing rocket and missile research. It wasn't until 1953, when atomic scientists came up with thermonuclear warheads light enough for first-generation intercontinental ballistic missiles (ICBMs), that the United States gave high priority to the military services' rocket and missile programs and began active development of the technology to be used both for war and the peaceful exploration of space. The Army, Navy, and Air Force each had missile programs in 1955; the Air Force, in fact, had three. Each of the services had a satellite program, too, and as it became feasible to consider actually sending a payload into space, the services found themselves in competition to be *the* U.S. effort. Vanguard, the Navy's collaborative, civilian-oriented effort with the National Academy of Sciences, was the choice for the first U.S. earth satellite program, and Americans were complacently anticipating its orbiting during the International Geophysical Year of 1957–58. Russia's Sputnik made it into space first; Vanguard blew up on the launching pad. The United States found itself unquestionably in second place in the new, high-stakes competition of the "space race."

Hindsight has shown that a lack of clear-cut goals, sound organizational structure, and appropriate funding were to blame for America's poor standing at that time. The confusion existing then is illustrated by the eleven major organizational changes that took place in the space program between 1953 and 1957, and the fact that by 1960 over 60 governmental groups were involved in space-related research. Writing in "The Cosmic Chase," a recent look at the politics of the space program, Richard Hutton observes that bureaucratic delays, interservice rivalries, governmental budget concerns, and "obsessive, constant efforts at reorganization" were a virtual guarantee that "a satellite program that was well within the capabilities of existing technology would not get off the ground until well after Sputnik had succeeded."[1]

Yet despite the program's late and slow start and the organizational disarray, it was only four years after the first Vanguard's ignominious demise that the United States flew a manned suborbital mission and had the temerity to announce plans to land a man on the moon by 1970. Moreover, it did just that on July 20, 1969. And the Vanguard project, despite its unfortunate beginning, did later succeed and proved a remarkably long-lived and useful satellite.

How the United States organized and carried out such an achievement in a relatively short time is the story of the National Aeronautics and Space Administration (NASA), a remarkably inventive organization that was itself an invention of the times and of a society that perceived, if somewhat suddenly, a need for it. The U.S. space program, as carried out by NASA, reflects in many ways the strengths of the nation that created it—its enthusiastic response to a challenge and a goal; a broad, humanistic view of inherent possibilities; a genius for applied technology and, as it turned out, for applied management as well.

It may well be that the current low status accorded the space program is equally a reflection of other qualities and preferences of the American people. Certainly, it is impossible to look at the social inventiveness that placed a man on the moon without asking why the nation has, in effect, retreated into unconcern over the outcome of its own achievement. It remains to be seen, at this juncture, how the society at large and its leadership will respond to the possibilities now so obvious as a result of the Space Shuttle *Columbia's* maiden voyage, successfully concluded on April 14, 1981.

Looking backward, however, it is clear that the U.S. space program, as directed by NASA, reflects a remarkable partnership of social and technological invention—innovation directed toward goals of considerable promise, both socially and economically. But again, the history of the program is a reminder that social and technological inventiveness must move together and complement each other. For if society does not perceive the need for technology, if it deems technology not sufficiently useful or too costly, or if it cannot absorb and adapt to technology— physically and economically *or* emotionally and intellectually—then social and technological invention will not move in tandem, with potentially disastrous results.

The genesis of NASA and the U.S. space program from a great number of programs and related bureaucracies was an impressive organizational feat, accomplished through the adept expansion and refinement of many existing management techniques. It reflected the American ability to move resources quickly where there is political agreement on a goal, and it was consistent with the American impulse to meet a challenge by whatever means possible.[2] NASA was what one foreign observer of the U.S. scientific scene saw as a favored type of American institution: "an ad hoc instrument" forged "in response to a challenge" and "generally in the form of an 'Agency' responsible directly to the President."[3] It brought together so many programs, disciplines, and talents and interwove them in such a way that it produced not only a number of technical achievements (miraculous in terms of the speed with which they occurred), but also a management achievement with profound social implications.

While NASA grew out of genuine national security concerns and a nationalistic urge to compete with the Russians, it also mirrored the strengths of the society that spawned it, and its program developed in a way that was consistent with human concerns. Despite its military potential and the concurrent development of military space capabilities, the American space program has retained a civilian emphasis for two decades. (There are indications this emphasis may undergo some change as new national priorities are established for the eighties.) Further, by focusing on major involvement by academia and industry (at the height of the program 90 percent of NASA work was contracted out; today the the figure is closer to 70 percent), NASA adhered to the "spirit of competition and the vigour of pluralism," characteristic of U.S. success in science and technology.[4]

NASA—A New Approach to Scientific/Technological Challenge

In terms of its ambitious objectives, the American commitment to the Saturn/Apollo program was possibly the greatest organized technological outreach ever undertaken by a nation, and in order to understand the magnitude of its successful accomplishment, one would have to start with the ingredients available in 1957. These were a level of technology largely related to military programs, an in-being organization devoted to scientific advances in aeronautics, and groups of scientists working— not necessarily together—on developing space exploration and experimentation concepts. When Sputnik came on the scene, a new element was added: a universal and unprecedented public clamor for U.S. achievement in this new domain. This high level of public support, as it happened, was finite in duration but at the time produced two off-shoots: a body of politicians supporting the objective of space achievement and the willingness of two presidents, Eisenhower and Kennedy, to respond to the expressed national need with a respectable national space program.

Public support and interest in Congress made federal funds available for large-scale financing, but though it was accepted that a considerable amount of money would be required, it was not clear how much was needed nor even how much could be usefully spent. First, it was important to determine the central orientation of U.S. space efforts. Scientists, generally, were pulling on one side for programs of true scientific value, while politicians and, presumably, a large segment of the public were pulling for more spectacular, high visibility objectives.

What *was* clear to most Americans, after Sputnik, was that space was a significant arena for national effort, and that the United States needed a sound, long-range space program. The Eisenhower Administration, Congress, and the interested public—particularly the scientific community—now sought the vehicle for implementation. The President's Science Advisory Committee (PSAC) led by Science Advisor Dr. James R. Killian of the Massachusetts Institute of Technology studied the issue and considered the various possible approaches, as did Congress. The program could, for example, have been centered in the National Advisory Committee for Aeronautics (NACA). NACA had an excellent reputation for its work in the engineering and design of aircraft; it had also worked on rocket propulsion and missiles. There were also those who felt the U.S. space program should originate within the armed services. A plan with much support involved combining the talents and facilities of the Department of Defense, NACA, the National Academy of Sciences, and the National Science Foundation, with NACA providing research and technical assistance and the Defense Department handling contracting for the development of hardware. Finally, there was interest in a broad-based program led by a new civilian agency. Many American scientists, through the American Rocket Society and the National Academy of Sciences, expressed a desire for a space program to meet broad cultural, scientific, and commercial objectives. This thinking was in tune with that of the Eisenhower Administration and the Advisory Committee that a distinction be made between civilian and military space efforts and that an establishment for directing the space program be scientifically oriented. President Eisenhower, in releasing the Advisory Committee's report, stressed his conviction ''that we and other nations have a great responsibility to promote the peaceful uses of space and to utilize the new knowledge obtainable from space science and technology for the benefit of all mankind.''[5]

The Advisory Committee itself cited four reasons for the advancement of space technology: (1) to satisfy the compelling urge of man to explore and discover, (2) to assure that space is not used to endanger national security, (3) to generate national prestige, and (4) to develop new opportunities for scientific observation and experiment which add to knowledge and understanding of the earth, the solar system, and the universe.

In Congress, where the legislative aspects of the space program cut across a broad range of jurisdictions, and new committee assignments were made to deal with the technicalities and the issues, legislators had various, often conflicting views of what the national emphasis in space should be. In the end, the legislative process, carried out under intense public pressure, resulted in a program beyond the scope envisioned by the agencies, particularly the Defense Department and NACA, most interested in focusing U.S. efforts in space.[6]

The National Aeronautics and Space Act of 1958 created NASA, a civilian space establishment, upon the structure of NACA, and while national security was an objective, it was not by any means the only or even the central consideration. When the new civilian agency opened its doors on October 1, 1958, a year after Sputnik, its objectives in aeronautics and the conquest of space were centered on:

- *Technology*—the development, operation, and improvement of aeronautics and space vehicles;

- *Political and national security interests*—the preservation of the role of the U.S. as a leader in aeronautical and space science and technology and an interchange of significant information with military agencies;
- *Science*—the expansion of human knowledge in the atmosphere and space and establishment of long-range studies with respect to the utilization of aeronautical and space activities;
- *Cooperation with other nations*—the exploration of space and the peaceful application of its results;
- *The most effective utilization of science and engineering resources*—the close cooperation among interested agencies in order to avoid duplication of effort, facilities and equipment.

The stress placed upon cooperation with other nations was of central importance because of the far-reaching consequences and implications of space exploration. Accomplishing the last goal, effective utilization of resources, was to prove a significant challenge that would give rise to the inventive management that characterized the American space program and resulted in NASA's success.

With NACA scientists, engineers, technicians and other personnel as the nucleus, NASA soon became a synthesis of many programs and incorporated a broad array of scientific and technical talents and disciplines. Earlier, the Administration had announced consolidation of DOD anti-missile and satellite activities in one organization, the Advanced Research Projects Agency (ARPA). Now various ARPA programs were transferred to NASA as were Air Force and Army projects. Also placed under NASA's wing were the IGY Project Vanguard, the scientific satellite program run by the Naval Research Laboratory; the Jet Propulsion Laboratory of the California Institute of Technology, which operated under federal contract; and the Army Ballistic Missile Agency in Huntsville, Alabama, where rocket expert Wernher von Braun was working and which became the NASA George C. Marshall Space Flight Center. This consolidation of whole blocks of skilled people with an already established teamwork proved an enormous strength for NASA. Each of these units had the ability to move on projects almost immediately. The directors of these field centers also provided the new national program with a strong cadre of mid-management leadership. From the first, coordination of these elements was a priority task for NASA top management, for the space agency at its outset was quite literally a pool of resources, talent, and knowledge. It was, in effect, an enormous research and development laboratory pulled together in fairly short order and staffed by people of widely dissimilar backgrounds, experience, and skills, bound together by some fairly monumental scientific and technical objectives and an exceedingly pressing agenda.

In retrospect, an amalgam of factors resulted in NASA's creation and allowed if not forced it to find the most effective means of achieving its goals. The public, upset over Sputnik, pressed Congress to act, and working in the spotlight Congress and the Administration could not simply settle for a patchwork of old programs. Once the public and large amounts of public money were involved, a bolder, new approach seemed to offer the best chance of success. A new agency provided the opportunity for an exciting administrative chemistry. The pressures of time and money had a catalytic effect, as did the daring of the mission itself. Imaginative, highly talented scientists and engineers were attracted to NASA, often willing to work at lower salaries than they might have found elsewhere for the unique research opportunities and chance to be part of an unprecedented scientific and technological endeavor.

In 1961, when the decision was made to extend NASA's reach to the moon, schedules became tighter than ever, the atmosphere more highly charged, and the need for creative management of available resources even greater.

Setting Goals, Public Support, and The Kennedy Decision

NASA first announced its goal of putting man into space, through Project Mercury, within a week after its establishment. By 1959, its planners had decided that the next logical step was a manned lunar landing and the agency included this goal in its first ten-year plan, a year after NASA came into being. That plan was an ambitious one, balancing scientific goals with engineering tasks. President Eisenhower deferred the idea of manned flight beyond Project Mercury; however, as the manned project attracted more attention and interest, it became the focus of NASA's program. Then on April 12, 1961, the Soviets once again stole the space show, sending Yuri Gagarin into orbital flight.

NASA had now lost its margin for error as it became clear what a failure—or a disaster, if a life were lost—would mean to the American people. Fortunately, Alan Shepard's May 5 flight less than a month later was successful, giving NASA the momentum and public confidence it needed.

Neither Nixon nor Kennedy, campaigning for the Presidency in 1960, was enormously pro-space. Nixon supported Eisenhower's policies, and Kennedy blamed the "missile gap" and the late U.S. start in space on the Eisenhower approach and policies, but he suggested nothing new.

After his election, however, Kennedy assigned Vice President Lyndon Johnson, who was strongly pro-space, to head the Space Council. Kennedy also set up an Ad Hoc Committee on Space to review the entire program and make recommendations. The Committee's report highlighted poor management in NASA and the military and manned space flight programs. It also criticized the overemphasis on manned flight suggesting, "We should find effective means to make people appreciate the cultural, public service, and military importance of space activities other than space travel."[7]

The Space Committee was very attuned to the public relations aspect of the focus on Mercury and the astronauts and concerned that major failures or disasters would endanger the entire program.

At that point world events intervened and overruled, in effect, the Committee's message. Gagarin's flight on April 12 was swiftly succeeded by the Bay of Pigs debacle in April 17, and President Kennedy was pressured to search for an immediate means of reaffirming his own leadership and building the country's sense of strength and confidence. The President and his advisors searched for a significant and challenging goal through which to turn the political tide. While there were other possibilities, such as the exploitation of the oceans or a massive developmental aid program to emerging countries, the space program, especially after Shepard's flight, offered the advantage of being in place and having a good chance of success. Having opted for space there were, again, several possible and highly visible goals—landing on the moon was the most spectacular and NASA believed it achievable based on technology capable of being developed within the decade.

The moon shot was Kennedy's choice, and although Kennedy was concerned that the space program would exploit technology in general for public benefit, the goal uppermost in the Administration's mind was to unite the American public.

"In a very real sense," Kennedy said, "it will not be one man going to the moon . . . it will be an entire nation."[8]

The elements of public support and political pressure were clearly critical to NASA's founding and its development and continued to play a large part in its orientation and fortunes through the sixties and seventies. From NASA's earliest days differences among even staunch supporters over program orientation were aired in the political arena with a primary focus on whether NASA's effort should be on manned or unmanned space exploration. And, of course, the scientists and engineers within NASA, or involved in its programs, tended to lean toward or support one argument or the other as it coincided with their views. NASA's orientation toward manned flight had political roots—certainly, political events led the Kennedy Administration to promote man's journey to the moon—but it was also related to the capabilities and interests of NASA planners and technical personnel.[9]

Through its relatively brief history, NASA has been criticized for playing politics and/or for playing them the wrong way. It has been accused of subverting valid scientific goals to those of the engineers and for publicizing its engineering feats to the detriment of other aspects of its program which have had far more impact on society and offer greater potential. Criticism of the emphasis on manned flight echoed through the seventies. At the same time, public reaction to the achievements of early scientific satellites waned, and public support for the space program that had been strong at first later flagged, despite widely beneficial technology transfer.[10,11] Since public outcry and support created the hothouse atmosphere in which NASA was established and the moon landing conceived, the phenomenon of public unconcern, if not outright rejection, of the space program in the seventies must also be noted.

After the moon landing in 1969, NASA lost momentum; public support was declining as evidenced by increasingly smaller budgets. Proposals for post-Apollo space stations or reusable spaceplanes were not received with much enthusiasm, and political support during the Nixon Administration was weak. Additionally, while technology had swept the United States forward into space, receptivity to technological change was not proceeding at the same pace. Social strains, perceptions of new responsibilities, Vietnam, and a growing awareness that scientific and technical advances may also have negative impacts created a climate of disenchantment. The space program itself generated worldwide interest in the environment; intense concern over ecology and resources could be dated from the Apollo photograph of the "precious and fragile earth."[12] Yet many environmental activists, had an anti-technological inclination, and the space program became a real target in the battle over social priorities for available funding. NASA found it had to scrap plans and live within significantly lower budgets, despite the impact of inflation. Much later, when the Shuttle was behind schedule and plagued by technical problems, a Carter Administration panel concluded that "while the shuttle's basic technology was sound, the program had suffered from years of economic malnutrition. Rather than asking for more money—and risking possible cancellation—NASA had taken engineering shortcuts and stretched out procurements, a practice that deferred (but increased) costs."[13]

Reduced public support and the budget stringency of the seventies caused a compromise of Shuttle capability, made it difficult to meet the schedule, and, some would argue, contributed to the technical difficulties.

NASA: A Social Invention on a Large Scale

The manner in which NASA came into being, as an act of will and determination of the American people, created not only the climate but an absolute need for the management solutions that came to constitute NASA itself. Inasmuch as an organization is a reflection of its role and manner of carrying out that role, NASA was the seminal invention of the American space program providing, as does any invention, "new capacity . . . to do a specific task."[14]

NASA's organizational inventiveness was forced by the need to work in the public spotlight and deal with the rapid expansion of a new program. Systems had to be developed for more efficient and responsible spending of large sums of public money and to coordinate work with other agencies with space interests, principally the Defense Department and Atomic Energy Commission. Much of NASA's technical mission had to be accomplished through contracting of work to industrial firms and universities, for while it developed a strong career staff and an outstanding technical competence within its own laboratories, the agency was also heavily dependent on industry for development and production of hardware and on universities for scientific contributions. This meant finding effective ways both to involve and coordinate industry and university talents and capabilities. From the start, NASA saw itself not as a technical resource agency as NACA had been, but as a management organization necessarily employing the most advanced communications and management techniques to direct the work of its network of federal laboratories, industrial contractors and university scientists.[15]

The nature of NASA's relationships with industry ranged from assuring contract compliance to offering the services of facilities and specialists and to stimulating the spinoff use of space knowledge and experience. In working with universities, NASA had to find ways to involve schools in space programs without competing directly for scientific and engineering talent. One approach was to rely on the university community as a scientific resource while permitting personnel to remain on campus. Another aspect of the NASA-university link involved development of specialized manpower, not only for NASA but for other national needs as well.

The variety and inventiveness of these relationships contributed much to the "chemistry" that created NASA and made it a stimulating atmosphere in which much was possible. There were also an openness and flexibility in the agency's approach that sprang from the size, daring, and the exploratory nature of its mission, which are reflected in the remarks of its first Administrator, Dr. T. Keith Glennan: "We are not an operating organization in the ordinary sense of that term. We do not expect to operate meteorological or communications systems. Our product is knowledge—new and fundamental knowledge—the techniques, processes and systems by means of which we acquire that knowledge."[16]

Important adjuncts of its complex mission and its other administrative goals were public information and education programs, which included the re-education when necessary of on-board employees. Another important element was NASA's motivational programs, which kept alive a sense of teamwork and commitment. Above all, the new agency had to create an environment conducive to scientific and technical creativity.

The space agency as it developed, as an inventive management concept, could not have come into being except in a society already favorably disposed to large, diverse, and complex organizations. Despite their aversion to centralized control, Americans have long been identified by their tendency to organize and create new organizational forms, e.g., the labor union, the corporation, the multiversity.[17] When a new goal required the focusing of vast national resources on the most complicated technical tasks yet embarked upon, an inventive management form was a natural and almost inevitable outgrowth. Advanced organizational and management skills were endemic, as was a large number of highly trained specialists. Science and technology had advanced to a level that made accomplishment of the envisioned tasks conceivable. Finally, an acceptance of the benefits of technology and the orientation of an optimistic nation toward science and technology and, indeed, toward world leadership combined to create the right environment for the "invention" of NASA.

Managing Technology Development—The Systems Approach

Federal support of knowledge-creating enterprises associated with public sector needs have focused on achieving rapid rates of technological advance in fields in which public interest transcends private incentives, as in the health field, on supporting research and development in weak industries, such as in agriculture earlier in this century, and on providing broad-scale support of basic R&D and science education.[18]

Space was clearly an area where private incentives were not equal to the enormity of either the task or the risk. Still, an early NASA decision to involve private industry (some 20,000 contractors in all) was a strength of the space program. Involving industry meant rejecting the Army's "arsenal" concept of creating government-owned and operated manufacturing facilities. The Air Force approach of contracting out to private companies was chosen instead. This permitted accomplishment of NASA's mission without an overly cumbersome bureaucracy. It also gave the new agency the benefit of industry-developed management techniques as well as research and development and production expertise. Other management input came from the Navy, when NASA adopted the Program Evaluation and Review Technique (PERT) developed to build the Polaris missile. PERT allowed each project manager to sort out the task that had to be done, when and in what order, and how long it would take for each. A PERT chart permitted one to see parallel paths of development and critical time points, to manage convergence of the paths in a finished product.[19]

A wide range of other management tools were developed and refined as the NASA space program progressed, facilitating the complex conceptual, planning, administrative, and evaluative tasks facing the agency and its contractors. The management approach that came to be identified with NASA had existed in various, often rudimentary forms, before NASA itself, but the nature and size of the agency's tasks forced it to develop an optimum method of managing many detailed, complex, and critical projects at the same time and making them move together toward a particular goal. The system did not develop overnight; nor did it proceed without difficulties. It was an evolutionary process, and, moreover, mangement concepts and techniques were developing along two tracks. It was the management of scientific and technical development that resulted in superb hardware for the hostile environment of space. But it was the management and direction of people and resources that allowed space science and technology to flourish and to mesh into sound, successful programs.

From the beginning, it was clear that good communications between the various elements of the organization were vital and a central planning system was needed to lay out possible options for the Administrator, who had ultimate responsibility for decisions. Within this centralized system, a planning approach developed involving steering, coordinating, and working groups led by senior managers and scientists from both NASA headquarters and the field centers. The idea was to overcome organizational barriers and reduce competition between centers, as well as duplication of effort. Senior NASA officials met weekly to discuss agency-wide policy problems as they related to program and operation in an effort to keep communications open.

A continuing problem in NASA's first years was the independence of the various centers around the country, an independence that was something of a legacy from the days of NACA when 8,000 field center employees were nearly independent of NACA's Washington staff.[20] Reorganizations were conducted in 1958, 1961, and 1963, the most important in 1961 when major program offices were created with clearly defined project management, and field centers were placed under the control of the offices within NASA that made the most use of them. The four major program offices were: Manned Space Flight, Space Sciences, Advanced Research and Technology, and Applications. Field center directors were provided with a direct line to the associate administrator heading the office to which they reported. Additionally, Program Evaluation and Review Techniques (PERT) were revised and reliability and quality assurance programs instituted. Industry and university efforts were now coordinated through new directorships set up to oversee the various programs, each reporting back to NASA Headquarters but maintaining its own identity and direction of its specific mission. Thus, during Apollo, the Houston Manned Spacecraft Center directed spacecraft development while the Marshall Space Flight Center focused on design and building of the Saturn rockets, and Kennedy Space Center managed launch operations. Research and development and construction at each center was consistent with that center's mission and was carried out within a framework of Phased Project Planning. Phased Project Planning involved:

- *Phase A—Preliminary Analysis.* Determination of achievability of an objective and identification of the most promising approaches and necessary project elements—from facilities to R&D.
- *Phase B—Definition.* Detailed study, comparative analysis, and preliminary design of system in order to choose the single most promising approach.
- *Phase C—Design.* Detailed definition of objectives and final project concept including system design with mockups and test articles of systems and subsystems to assure hardware was within state-of-the-art, that schedules and resources estimates were realistic, and that definitive contracts could be negotiated. Alternate and backup systems requirements were also detailed.
- *Phase D—Development Operations.* Final hardware design and development, as well as production, test and project operations.[21]

Coordinating the network of field centers was not an easy task as thousands of events happening in as many places had to occur at the right time. Additionally, many of these events involved research and development: the answers had still to be found. A breakdown in progress at any point could halt progress in major program segments and an unexpected advance, on the other hand, could require major adjustments throughout hundreds of management interfaces.

NASA and its contractors developed management techniques that provided the efficiency and flexibility to rapidly identify problems at any level before they could affect overall progress. They applied them through the basic management structure of the Office Of Manned Space Flight at NASA Headquarters in Washington and the three centers primarily concerned with Gemini/Apollo: the Manned Spacecraft Center in Houston, Texas; the Marshall Space Flight Center in Huntsville, Alabama; and the John F. Kennedy Space Center, Cape Kennedy, Florida.

The management structure of contractors and subcontractors was below this basic organization and here scheduling and review procedures were uniform and fully integrated. Scheduling depended upon technical progress, funding, and manpower. The Centers and contractors directly responsible to NASA Headquarters prepared their schedules after analyzing reports from other contractors and subcontractors. This information was then submitted for evaluation and decision to the Office of Manned Space Flight. The scheduling procedures provided current information on the status of hardware development and production at all levels for the use of management at any level.

Communication of information so that all technical relationships were clear and understood was a particular management challenge. At times, these relationships had to be strengthened, as after the Apollo fire in 1967. To make certain that its potential problem warning systems were functioning effectively, NASA established an Office of Organization and Management at headquarters to coordinate a number of offices scattered throughout the organization in an earlier decentralization move. The new office, headed by an associate administrator, was to develop methods for review and approval of all major management actions and resources, to see that policies and practices were followed, and to assure that requirements of Congress and other federal agencies were met. An Office of Assistant Administrator for Special Contracts Negotiation and Review was established within the Office of Organization and Management to give particular attention to procurement actions.[22]

While NASA headquarters organized the upper-level supervisory elements of the space program and coordinated the various parts of its far-flung team, the job of developing the hardware and making it work was left primarily to industry. Industry performed this function, using and refining the systems approach which had earlier been applied in some form in large-scale commercial and construction projects and in federally-supported R&D programs. Antecedents of the systems method were seen in merchandising and administration programs of R. H. Macy & Company in the 1930s and the Radio Corporation of America's planning for television broadcasting. An emerging systems approach was used in England during World War II in the development of radar.[23]

The U.S. aerospace industry had experience in the systems approach dating from the early fifties and the development of supersonic fighters. Production of these aircraft, of greater complexity than their predecessors, required greater inter-system, inter-company coordination than ever before. These management techniques were then expanded under the Weapons System Concept in the ballistic missile program of the late fifties and early sixties and the Air Force introduced the Weapons System Concept in the development of the B-36 bomber. In essence, its purpose was to insure that all components of a complete system met all performance and reliability requirements at a given time at lowest possible cost. It provided several contractual approaches to managing the procurement of portions of the system, monitoring the progress of work, and integrating the various elements into the final system.

Industry used the systems approach as it worked to develop and produce hardware for space, but its contribution went beyond scientific and technology advances. It included management techniques such as value engineering, program evaluation and review, and progressive procurement programs—all important in the search for better and cost effective ways of accomplishing a project.

NASA's top management, of course, was responsible for a total systems integration—for assuring that all of the systems, of whatever level of complexity, fit successfully together. Though developed at different times and by different companies, each technical system had to work perfectly by itself and when integrated into the whole. Furthermore, each management unit below headquarters level was also a system, included subsystems that had to work together, and had to mesh its efforts, finally, with those of the other units.

The way in which the agency approached its management task was influenced by the background and traditions of the research and development organizations that had been merged into NASA. These included: outstanding records of scientific and engineering achievement, a tradition of close cooperation with industry and universities, and a history of utilizing the latest technological innovations for both technical and management problems. Personnel in those organizations had also been highly career-oriented. These considerations influenced NASA's management plan which emphasized technical excellence, top management involvement in planning and implementation, strong in-house technical competence, project management, and in-depth monitoring of contractors.[24]

Because Apollo was primarily a technical task, the top mangement team included technical experts, offering sensitivity to the needs of technical personnel while also giving them status and wide decision-making powers at all levels. At the same time, top management had to be involved not only in the commitment of resources and setting of policy, but in the process of execution. Since going to the moon had nothing of the routine about it, top management had to have a hand in how it was to be brought off. A strong in-house technical competence was necessary so that NASA could judge the capabilities and oversee the performance of the industry and university community which was doing 90 percent of the research and development.[25]

The project was the central focus in NASA's organizational structure with a manager responsible for all activities and objectives within time and cost limits. It was the project manager's task to motivate and integrate the efforts of staff with those of other specialists. The usefulness of project organization was that it permitted contraction or expansion as the project's overall mission was accomplished or enlarged.[26]

Finally, in-depth monitoring of outside performance was necessitated by the risk factor involved in the NASA mission and by the absolute need to succeed.[27]

Overall, NASA's approach facilitated a smooth relationship of experts from many disciplines and let the best, most cost-effective ideas come to the fore. It provided managers with the tools to handle many constantly changing factors including money, time, people, and environmental and social conditions.

The technical management teamwork that NASA represented was possible because the American society demanded it in fulfillment of a national goal and could support it with both the necessary technical and financial resources. In that sense, it was social invention, and its evolving management structure and methods were socially innovative, permitting both the technical breakthroughs and the technical, economic, and social innovation that continues even today. The nature of technological change in this century and the urgency of not only discovery but of innovation were already such that the complexity, interdisciplinary nature, and requisite huge investment in materials and equipment in certain areas necessitated organized activity of the kind epitomized by NASA. While the lone inventor—the scientist, technician, or tinkerer—continues to make remarkable contributions, there is no denying that technology in the twentieth century depends on the professionalization, specialization, and systematic application of science to technology made possible by the team approach.

The space program systems approach has been termed "genuine innovation . . . generated by twentieth century trends in technology: the linking of technology and science, the development of the systematic discipline of research, and innovation. The systems approach is, in fact, a measure of our newly found technological capacity. Earlier ages could visualize systems but they lacked the technological means to realize such visions."[28]

Since the new management methodology grew out of the convergence of trends in science and technology, it necessarily made everything move more swiftly. It created new generic concepts—"materials" rather than steel, glass, concrete, for example—permitting design with an end use and desired property characteristics clearly in mind. This revolutionary approach, devolving from the systems concept, has profoundly affected other technologies and will increasingly affect mankind.[29]

NASA's Contribution to Social, Technological and Economic Development

The physical manifestations of NASA's inventiveness will have continuing impact. But, in addition, the very means by which their creation was managed has given rise to useful and far-reaching approaches to dealing with other technical problems and with the broader problems of society. These techniques can be used in applying new technology to the solution of socio/economic problems, including those that center on society's adaptation to technological change. Management approaches for directing massive and sophisticated projects are in themselves a spinoff of the space program and perhaps, in the long run, will prove among the most important.

Systems management is most easily applicable to technological enterprises; nonetheless, it has been used effectively in socio-economic endeavors, particularly when they involve application of technology to social needs. The development of transportation systems and delivery of health care services are excellent examples of the transfer of these management skills.[30] A recent commercial example of the transfer of project management and systems analysis skills centers on the planning, design, development, and management of major agricultural programs in the United States and developing countries; the company involved developed its expertise through work with NASA and other federal agencies.

It has been suggested that the NASA model itself would work in energy technology development, but necessary ingredients such as a greater sense of urgency on the part of the public, greater government support, and a more effective management team than yet assembled by the Department of Energy seem to be missing. It is far more uncertain whether the model would work in a crash cancer research program or any other in which there are so many unknowns of a scientific rather than technological nature. Moreover, the NASA management approach did not work as well in the Space Shuttle program as in Apollo, indicating perhaps that the effectiveness of the management team, as well as adequate financial support, is a key factor.

The human element is always critical, in fact, whether in the development of technology, the application of technology, or in any other realm of factor or resource management. Human judgment *must* be the overriding factor superimposed on quantitative data. And when human actions are the target, as in social management programs, methods and techniques that work well in technology development may have more limited success. The systems approach is still a powerful decision-making tool and is useful in building public understanding of options and variables in any issue. It has been said that the systems method's real potential for solving social problems is that it encourages action, not "crisis action," but the habitual use of an approach that provides a "steady flow of clues to predict and forestall cataclysmic effects of inaction."[31]

Space technology transfer has taken a wide variety of forms, both everyday and esoteric. Space age "hard" technology is in evidence in our daily lives from a multiplicity of new, or improved, household products to major advances in health services. Most importantly, space technology combined with management skills is expanding capabilities in far-reaching ways; for example, the ability to forecast weather, to harness new energy sources, and to locate and chart earth's resources will permit us to better manage resources and adapt to technological change. But to date the possibilities inherent in the advanced communications of the space age have had, and undoubtedly will continue to have, the greatest societal impact. Communications satellites and advanced computer systems made possible by microcircuitry have led to genuine social change through instantaneous communications and the capability for real-time decision-making.

Not the least of the benefits accruing from the space program has been an economic ripple effect made possible by the market-expanding introduction of new products, new scientific and industrial processes, new methods of handling materials and organizing production, and the new techniques of planning and management. The resulting economic growth further assisted development of the scientific/technological base and provided stimulus, through increased productivity and reduced costs, for a number of major industries. New products and new markets have meant new jobs as well.

The pressures of world population increase and the depletion of natural resources have themselves forced a realignment of economic capabilities among nations. Space-generated technology has helped the United States and other advanced industrial nations to adjust and adapt. Other developed countries, as well as the developing nations of the Third World, have expanded production and trading capabilities by building, in many instances, upon the advantage of lower labor costs.

The balance of comparative advantage has shifted, and advanced technology, some of it developed in the space program or through NASA's continuing aeronautical research and technology development, now serves as a major advantage for the leading industrial countries in a far more competitive world.

A further and no less important offshoot of NASA aeronautical and space programs has been development of a resource implicit in high technology: a highly skilled workforce. The spread of knowledge, rise of skill level, and development of high standards of specialization and professionalism that were concomitants of the NASA program will be important assets as society strives to plan for and adapt technology for broad social benefit.

Notes

1. Richard Hutton, *The Cosmic Chase* (New York: New American Library, 1981), p. 45.
2. Théo Lefèvre, "Reflections on Science and Policy in the United States," *Reviews of National Science Policy: United States* (Paris: Organisation for Economic Cooperation and Development, 1968), p. 361.
3. Ibid., p. 358.
4. Organisation for Economic Cooperation and Development, *Reviews of National Science Policy: United States* (Paris: 1968), p. 95.
5. *Historical Sketch of NASA,* prepared by the NASA Historical Staff (Washington, D.C.: National Aeronautics and Space Administration, 1965), p. 6.
6. Richard Hirsch and Joseph Trento, *The National Aeronautics and Space Administration* (New York: Praeger, 1973), p. 28.
7. Hutton, *Cosmic Chase,* p. 71.
8. Quoted in Hutton, *Cosmic Chase,* p. 75.
9. John Noble Wilford, "Riding High," *The Wilson Quarterly,* Autumn 1980, no. 698, April 21, 1981, Special Edition, pp. 3–4.
10. Hutton, *Cosmic Chase,* p. 13.
11. *Outlook for Space,* Report to the NASA Administrator by the Outlook for Space Study Group (Washington, D.C.: National Aeronautics and Space Administration, 1976), p. 17.
12. L. B. Taylor, Jr., *For All Mankind: America's Space Programs of the 1970s and Beyond* (New York: E. P. Dutton, 1974), pp. 135–136.
13. Wilford, "Riding High," p. 7.
14. Peter F. Drucker, *Technology, Management and Society* (New York: Harper & Row, 1958), p. 67.
15. Hirsch and Trento, *National Aeronautics and Space Administration,* p. 41.
16. Alfred Rosenthal, *Venture Into Space: Early Years of Goddard Space Flight Center,* NASA Center History Series (Washington, D.C.: National Aeronautics and Space Administration, 1968), p. 39.
17. Fremont E. Kast and James E. Rosenzweig, *Organization and Management: A Systems and Contingency Approach,* 3rd ed. (New York: McGraw-Hill, 1979), p. 131.
18. Richard R. Nelson, Merton J. Peck, and Edward D. Kalachek, *Technology, Economic*

Growth and Public Policy (Washington, D.C.: The Brookings Institution, 1967), pp. 163–164.
19. Wilford, *"Riding High,"* p. 4.
20. Hirsch and Trento, *National Aeronautics and Space Administration*, p. 42.
21. Hutton, *Cosmic Chase*, p. 161.
22. Hirsch and Trento, *National Aeronautics and Space Administration*, pp. 44–45.
23. Frederick I. Ordway, Carsbie C. Adams, and Mitchell R. Sharpe, *Dividends From Space* (New York: Thomas Y. Crowell Company, 1971), p. 64.
24. Michael J. Vaccaro, *A Systems Approach to the Management of Large Projects: Review of NASA Experience with Societal Implications*, National Aeronautics and Space Administration report (Greenbelt, Maryland, 1973), p. 3–6.
25. Ibid.
26. Ibid.
27. Ibid.
28. P. Drucker, *Technology, Management and Society*, pp. 70–71.
29. Ibid., p. 71.
30. Vaccaro, *Systems Approach to Management of Large Projects*, p. 6.
31. Simon Ramo, *Cure for Chaos: Fresh Solutions to Social Problems Through the Systems Approach* (New York: David McKay Company, 1969), p. 113.

chapter ten

ABSORPTION AND ADAPTATION: JAPANESE INVENTIVENESS IN TECHNOLOGICAL DEVELOPMENT

Tetsunori Koizumi

"**A**s first perceived, the outward strangeness of things in Japan produces a queer thrill impossible to describe—a feeling of weirdness which comes to us only with the perception of the totally unfamiliar."[1] This is how Lafcadio Hearn recounts his impression of his first encounter with Japan. When he wrote these words in *Japan: An Attempt at Interpretation* in 1904, Hearn had been in Japan for some fourteen years, married a Japanese lady, traveled widely in Japan, and written extensively on Japan under his adopted Japanese name, Koizumi Yakumo. However, further acquaintance with Japan which he had gained during these years did nothing to diminish the sense of strangeness he had experienced on his first encounter: "Further acquaintance with this fantastic world will in nowise diminish the sense of strangeness evoked by the first vision of it."[2]

The "sense of strangeness" which Hearn experienced can be partly attributed to the fact that his contact was with Meiji Japan, a few decades after the Meiji Restoration had unveiled this "fantastic world" to the West, when the life of the Japanese was still very much influenced by the legacy of Tokugawa Japan. But Hearn also witnessed the efforts made by the Japanese to modernize their country and correctly foresaw Japan's potential to develop into an industrial power to be reckoned with. To the extent that the Japanese have been successful in their efforts, we would expect the sense of strangeness Japan evokes to Westerners to have diminished, if not completely dissipated. For the process of modernization and industrialization for Japan since 1868 has, at the same time, been a process of "Westernization" as the Japanese have tried to catch up with the advanced countries of the West.

As it is, the sense of strangeness Japan evokes to Westerners appears to be experiencing somewhat of a resurrection long after the Japanese have successfully accomplished their task of catching up with the advanced countries of the West. Within the business community where the resurrection is most pronounced, Western industrialists seem to be puzzled by Japan's industrialization which owes its success to the ways of doing things which do not conform with their conventional wisdom. Whether the Japanese ways of doing things are conventional or not, one thing is certain: Not only have the Japanese gradually washed away the competitive advantage of the Western economies, they have also surged ahead in some vital areas of technology. In fact, some studies suggest that Japan now has a sizable advantage in overall technology over the United States and, therefore, an even greater advantage over other Western countries.[3] It is not surprising, then, if the sense of strangeness, which used to be mixed with wonder and amazement, is now sometimes tinged with worry and suspicion.

What is the source of this sense of strangeness which is still expressed by Westerners towards Japan? What is special about Japan's industrialization other than the fact that she is the first non-Western country to join the select club of industrial powers? What is the secret behind Japan's success in technological development? Answers to these questions hold the key to understanding the role of social inventiveness in technological development not only in Japan but also in other countries. This is so because technological development, although it primarily depends on the fruits of science and technology, also depends on the ways in which the efforts of the population in general are socially mobilized. And it is here that technological development becomes a matter of "social" inventiveness in that it depends on how effectively the fruits of the phylogenetic learning of a nation are exploited.

Is there, then, anything unique about the social inventiveness of the Japanese in technological development? To the extent that the fruits of phylogenetic learning are at stake, what Hearn observed at the turn of the century may still hold a key to understanding the nature of Japanese inventiveness in technological development: "Tools are of surprising shapes, and are handled after surprising methods: the blacksmith squats at his anvil, wielding a hammer such as no Western smith could use without long practice; the carpenter pulls, instead of pushing, his extraordinary plane and saw."[4] Why does the Japanese carpenter pull, instead of pushing, his plane and saw? If we are interested simply in what the carpenter makes, we may not pay much attention to the way the carpenter uses his tools—after all, a desk is a desk is a desk! If, however, we detect a significant difference in the quality of the product made in Japan and elsewhere, then we begin to suspect that there is perhaps something fundamental about the way the Japanese carpenter uses his tools. The purpose of this chapter is to analyze if this simple act of the Japanese carpenter who, unlike a Westerner, pulls his plane and saw, indeed symbolizes what is fundamental about the nature of Japanese inventiveness in technological development which has been acquired during the process of their phylogenetic learning as a nation.

The Legacy of Tokugawa Japan

It is tempting to associate the image of peace and stagnation with Tokugawa Japan considering that the nation was ruled for two and a half centuries by hereditary feudal lords under the policy of national seclusion. There was indeed peace compared with the turbulent period of civil wars which preceded it and lasted more than a century until Ieyasu firmly established his hegemony over other feudal lords at the battle of Sekigahara in 1600. And there was stagnation too, especially in the second half of the Tokugawa period, marked by stationary population caused by famines which followed crop failures and by infanticide which was practiced to relieve economic misery. However, to characterize the whole Tokugawa period as the period of peace and stagnation, as has been done by many historians,[5] would be a serious over-simplification. Beneath the calm surface which gave the deceptive appearance of peace and stagnation there was definitely a strong current of social change engulfing all aspects of Japanese life. Indeed, without due consideration of the nature of social change in Tokugawa Japan, it would be difficult to understand the promptness with which the whole nation embarked on the task of modernizing Japan since 1868 and the remarkable story of her success in industrialization, though interrupted by a few waves of depressions along the way and a total wreckage at the end of World War II. Let us, therefore, begin our analysis with a discussion of the nature of social change in Tokugawa Japan.

Tokugawa Japan, politically speaking, was under the feudal reign of the Tokugawa family. As such, it was a variation of "clan feudalism" which had been in existence in Japan throughout history since as early as the fourth century. What distinguished Tokugawa feudalism from the earlier forms, however, is that its reign covered practically the entire archipelago, with the exception of Hokkaido which was yet to be cultivated. That the political unity established by the Tokugawa family was on a truly national scale was no doubt responsible for the "apparent" solidarity of Tokugawa feudalism. Yet a closer scrutiny of the Tokugawa polity easily reveals a structural weakness, inherent from the very beginning.

The feudal structure of Tokugawa Japan was centered around the Shogunate which sat at the top of a caste system consisting of samurais, farmers, artisans, and merchants. The Shogunate was located in Edo with *gosanke* (three exalted families) strategically placed in Mito, Owari (Nagoya), and Wakayama. The reign of the Shogunate, however, did not extend uniformly over the whole nation, for the job of administering local fiefs was entrusted to *kokushus* (local lords). The upshot was that each *kokushu* could exercise his own discretion in administering the affairs of his fief. This was especially true with *tozamas* (outside lords) whose fiefs were literally located far away from Edo. In fact, the rebellion against the Shogunate which culminated in the Meiji Restoration came from three of these outside fiefs: Satsuma of Kyūshū, Tosa of Shikoku, and Chōshū of western Honshū.

There is no question that the rebellion against the Shogunate was politically motivated to bring an end to the system which was becoming more and more ineffective in maintaining national unity. But the ineffective management of the economy by the Shogunate also contributed to the breakdown of Tokugawa feudalism. In the realm of economics, as in politics, the role of the Shogunate was limited from the very beginning. This is because the revenue of the Shogunate came almost exclusively from land taxes. To be sure, there was some revenue from foreign trade; but the amount dwindled quickly as the policy of national seclusion went officially into effect in 1641 with the confinement of Dutch traders to Nagasaki. As the revenue from trade dwindled, the the Shogunate had to depend more and more on land taxes for its revenue. But there was, of course, a limit beyond which farmers could not be squeezed. To make things worse, only about 15 percent of the land was directly held by the Shogunate and another 10 percent by *hatamotos* who, as the retainers of the Shogunate, directly served Shogun in Edo. The rest of the land, with a minimal amount set aside for the Imperial Court in Kyoto and certain temples and shrines, was under the direct control of the local lords.

As for the local lord, the job of keeping his retainers well provided for was not an easy one. The problem was further complicated by the hierarchical structure which existed within the samurai class. While the upper samurais were fairly well provided for, the lower samurais had to struggle to make ends meet by seeking secondary sources of income. Some of them, taking advantage of the skills they acquired, became merchants. In fact, the lower samurais from Satsuma, Tosa, and Chōshū who rebelled against the Shogunate and became the leaders of the new Meiji Government were clever businessmen in addition to being political idealists. Thus the solidarity of Tokugawa feudalism was undermined by the ruling samurai class who, as privileged rentiers, lacked the kind of dedication and entrepreneurship needed for a successful management of the economy.

If the ruling samurai class was not to be counted on, the accumulation of the wealth of the nation had to depend on the wits and inventiveness of the lower echelons of the society. There is indeed a bit of irony here, for the "apparent" political stability which the Shogunate provided the nation for as long as it lasted made it possible to cultivate all kinds of production activities at the local level, ranging from food and clothing to metal and handicraft. And the ineffectiveness of the Shogunate in managing a national economy led to the development of a market economy on a national scale as the thriving local productions by lower samurais and various guild organizations gradually expanded the network of commodity transactions. With the expansion of the network of transactions, money emerged as a means of exchange. Even the upper samurais contributed to the development of a monetary economy as they used cash on their trips to and from Edo which were part of their feudal obligations to the Shogunate.

There was indeed movement of both people and commodities in Tokugawa Japan. The natural result was the development of a transportation network. As is well recognized by economists, a transportation network as a form of social overhead capital plays a strategic role in economic development. It is perhaps relevant here to note that the Roman Empire lacked the kind of transportation network, including merchant marines, that was needed to maintain economic integration over the whole domain. Would it be safe to argue, by analogy, that the Tokugawa Shogunate made the same mistake when it decided to adopt the policy of national seclusion?

We know for certain that the revenue from trade for the Shogunate and the nation dwindled to an insignificant amount. Had there not been movement of cargos between Edo and Osaka on the sea, the isolationist policy could well have cost the Shogunate the invaluable reservoir of know-how in shipbuilding and navigation that it inherited from the earlier periods. Japan had indeed enjoyed a long history of shipbuilding and navigation, dating back to the seventh and eighth centuries when the official missions of the Court visited Sui and T'ang China. In view of such a tradition, the appearance of detailed navigational charts which included not only China but also Borneo and Makassa in the Muromachi period was no accident. In fact, the Ashikaga Shogunate had built enormous vessels to be used for trade with Ming China. In the fifteenth and sixteenth centuries Japan was an equal match for England and Spain in terms of expertise in shipbuilding and navigation. The isolationist policy of the Tokugawa Shogunate was definitely a serious setback to this seafaring tradition of the Japanese. But thanks to the expansion of domestic trade, this proud tradition survived Tokugawa feudalism without total extinction.

If the robustness of economic activities was any indication, Tokugawa Japan was actually a period of remarkable social change. It was, of course, a period of population increase, at least in the first half of the period. When the first census was taken in 1721, the population of Japan stood at about 30 million, well over a 50 percent increase over that in 1600, and had already exceeded the population of any European country.[6] It was also a period of urbanization. Needless to say, there had been urban development in earlier periods, especially around major religious and commercial centers. In the Tokugawa period we saw the development of *jōkamachi* (castle town) as the headquarter of a feudal lord. Edo, as the headquarter of the Shogunate, was the largest of such castle towns—indeed the largest city of the world at the time when the first census was taken with the population exceeding one million. Besides Edo, such castle towns as Nagoya and Kanazawa boasted the population of about 100,000 which was comparable to such European cities as Amsterdam and Rome.

Castle towns attracted people of all ranks, from samurais down to farmers, artisans, merchants, and prostitutes who were officially outside of the caste system. Farmers came to town to seek menial jobs required by samurais, artisans to take advantage of their skills, and merchants to supply materials for samurais. As for prostitutes, they came to town to "entertain," for most of them belonged to traveling Kabuki troupes. Channels for widespread social change were thus opened up for *chōnin* (towns people) as these common people could now interact directly with local lords in these castle towns.

All in all, Tokugawa Japan was indeed a period of remarkable social change. Although the end of national seclusion came dramatically in the form of an external threat with the arrival of Commodore Perry in 1853, there was also an internal threat which was gradually undermining the foundation of Tokugawa feudalism. Contrary to the image of peace and stagnation which Tokugawa Japan may evoke with its policy of national seclusion and feudalistic social structure, its greatest legacy may indeed have been the inventive and innovative spirit, not of the ruling samurai class, but of the common people who were the real actors of its social drama.

The Role of Education

Cultivation of both mind and body was part and parcel of samurai training. And Ieyasu, though himself a military man whose training was done mostly on a horseback, was acutely aware of the importance of education. In fact, the house testament of the Imagawa family in whose custody Ieyasu spent his earlier years said, ''He who does not know the Way of *Bun* (the literary arts) can never ultimately gain victory in the Way of *Bu* (the military arts).''[7] Thus Ieyasu willingly subsidized the printing of books, financed the construction of libraries, and encouraged study not necessarily for samurais but also for the populace. Ieyasu's effort in restoring cultural life was duplicated by other feudal lords.

A renewed interest in cultural life was, of course, made possible when Ieyasu's campaigns brought an end to a prolonged period of civil wars. This does not mean, however, that there were no cultural activities during these war-ridden years. There were, in fact, sporadic outbursts of cultural activities carried out by special groups of artists who as members of such schools as Kano, Tosa, and Rimpa, had the financial backing of feudal lords and merchants. When Ieyasu completed his campaigns, however, cultural life for the average Japanese was virtually non-existent. In fact, with the exception of priests, scholars, and doctors, the population was largely illiterate. All this was transformed dramatically by the time the Tokugawa Shogunate gave way to the Meiji Government, for the literacy rate around 1870 was probably well over 70 percent, a figure which compared favorably with other advanced countries of the West.[8] This remarkable achievement in education is yet another evidence of the kind of social change that took place in Tokugawa Japan.

What is most significant about the legacy of Tokugawa education is that it consisted mainly of general education: reading, writing, and arithmetic. Although the exact chain of causality is difficult to establish, it is well recognized by economic historians that there is an important correlation between the potential for economic development and the level of general education as measured, for example, by the literacy rate.[9] It may be illuminating, therefore, to look back on Japan's economic development since 1868 from this perspective.

There are many reasons why general education plays a key role in economic and technological development. For one thing, general education cultivates self-discipline. In the case of Tokugawa education, this point was explicit because education, especially of the samurai class, emphasized the development of one's moral character. But there is no need for this point to be explicitly stated as the purpose of education, for general education helps to develop self-discipline because it involves constant repetition and memorizing. It is one thing to preach the virtue of work but quite another to teach a work ethic. To the extent that self-discipline is needed in performing simple tasks, it would not be far-fetched to trace the origin of the work ethic of Japanese workers today to the legacy of Tokugawa education.

There is also the role of general education in facilitating the diffusion of information among the general public. In fact, the communication between government and people is but one aspect of such a information-diffusion process. Without due consideration of this aspect of general education, it would be difficult to understand how the whole nation was able to respond to the initiative of the Meiji Government in the early stages of Japan's industrialization. It is also worth noting that Japan's industrial community has inherited this tradition of constant communication between the leader and the led in the form of consensus decision making.

General education also plays an important role in taking advantage of imported technologies. This point is especially important in relation to the Gershenkron hypothesis in economic development, which asserts that a late starter has an advantage over an early starter in that the former has access to all the up-to-date technologies developed by the latter.[10] A late starter, however, would not be able to exploit *his* advantage unless *his* population has the capacity to absorb and adapt imported technologies. Again in this connection, Japan has greatly benefited from the level of general education achieved by her workers. The practice of having inventive and innovative ideas come from the bottom up, which is inherent in the operation of quality control groups found in Japanese manufacturing firms today, would be impractical unless all the workers involved have attained a certain uniform level of general education.

Further, general education provides a stepping stone for potential leaders of industry and trade. Although much can be said against a large bureaucratic organization, it is worth noting that throughout the process of Japan's industrialization the government bureaucracy has been able to provide the necessary leadership thanks to the best talents assembled from the general public. In this regard, the role of bureaucracy in Japan is quite analogous to that in China which had the open system of entrance examinations as far back as the second century B.C. There is indeed something to the Oriental conviction, "the sword might win but only the logos can keep," for the waves of economic instability threatened the steady path to Japan's industrialization whenever the military machine held sway over the bureaucratic machine.

General education, as a form of human capital and public investment, is something whose return is not expected to be reaped in the short run. It is indeed a tribute to the insight of Tokugawa educators that Japan has been able to benefit from the legacy of Tokugawa education in her effort to attain modernity. Needless to say, education is by definition a two-way communication between the educator and the educated. A success in education would not have been possible if not for the willingness to accept education on the part of the general public. Personal advancement can, of course, be an important motivation to education, but the willingness to acquire knowledge for its own sake is equally important. Although cultural life was at a low ebb during the civil war years, the Japanese were never lacking in their willingness to acquire knowledge. Thus St. Francis Xavier, although he may have failed in his mission to convert Japan to Christianity, had quite correctly perceived the innate curiosity, the *will to know*, of the Japanese as early as in 1549:

> They are a well-meaning people and very sociable and anxious to learn. They take pleasure in hearing of the things of God, especially such as they can understand, and they have no idols made in the shape of beasts, but believe in men of ancient times who, as far as I can make out, lived as philosophers. Many Japanese adore the sun and others the moon. They like to be appealed to on rational grounds, and are ready to agree that what reason vindicates is right.[11]

The edict of national seclusion issued by the Shogunate, which included the ban on the teaching of Christianity, may have prevented the influx of ideas from foreign countries, but it certainly did not succeed in suppressing the innate curiosity, *the will to know*, of the Japanese. In fact, even the Shogunate were not immune from curiosity about foreign countries as the stories of foreign people and things were brought back to Japan by those who dared to risk their lives against the edict. Thus in 1793, Shogun Ienari summoned one Daikokuya Mitsudayū to his palace in Edo to hear directly from him what he had seen and heard in Russia. Mitsudayū, after a storm had taken his ship and crew to Amchitka in Aleutian Islands in 1783, had to spend the next ten years traveling from one place to another which included his trip to St. Petersburg to have an audience with Catherine the Great. Shogun listened intently to Mitsudayū as he told him of the Empress' interest in trade with Japan.[12]

There is another fantastic adventure story of Nakahama (John) Manjiro. Manjiro was an unknown fisherman from Tosa when he went out to sea one day in January of 1840 in his small fishing boat and was caught in a storm. He was saved by Captain Whitfield of an American whaler and accompanied him to America. His adventure for the next ten years would take him to Honolulu, Massachusetts, Manila, California, and finally back to Ryūkyū in January of 1851. Because of his violation of the edict, Manjiro had to spend nine more months in confinement. When he was finally released and sent back to his native Tosa, he went on to write a book on his adventure in which he correctly predicted the imminent arrival of Commodore Perry and his Black Ship, for he had overheard captains of American whalers complaining about the isolationist policy of the Tokugawa Shogunate as they desperately wanted to use Japanese ports to refuel their ships and procure materials for their crew.

If we are to sum up the nature of Japanese inventiveness in technological development, we may perhaps choose two words: absorption and adaptation. We are choosing these two words not necessarily because absorption and adaptation of imported technologies have played key roles in Japan's industrialization since 1868. These words sum up the nature of Japanese inventiveness in general which is much older than the story of Japan's industrialization or the legacy of Tokugawa education. As the fundamental traits of the Japanese mind, absorption and adaptation have played key roles in the evolution of Japanese culture whose spiritual root can be traced back to animism. Animism, as it has evolved into the spirit of Japanese culture, is not, however, to be regarded as a primitive form of religion. Rather, it was a world view, in which man is seen as an integral part of nature in that his presence in nature is linked with other things through a common element called, "spirit." In such a conception of the world there is a place for everything—the tree, the herb, the grass, the word, and certainly the spirit of one's ancestor. Psychologically speaking, the spirit of Japanese culture can be regarded as an expression of the all-embracing and all-engulfing aspect of the Great Mother archetype.[13]

It is this animistic world view, as it has evolved into the spirit of Japanese culture, that gives rise to absorption and adaptation as the two fundamental traits of the Japanese mind. This explains, for example, why the Japanese have managed to maintain both conservative and innovative spirits, the conservative spirit probably originating in ancestor worship and the innovative spirit in the will to know. Such a mixture of conservatism and innovation has certainly played an essential role in technological development as the Japanese have had to absorb technologies imported from abroad and adapt them to suit their domestic conditions. To the extent that absorption and adaptation reflect the spirit of Japanese culture, the Japanese experience since 1868 suggests the existence of an inseparable linkage between cultural evolution and technological development in a country's economic development. But has Japan been unique in this? Or can something more general be said about the linkage between technology and culture? Answers to these questions hold further keys to understanding the role of social inventiveness in technological development.

Technology and Japanese Culture

Technological development, broadly interpreted as a process of making improvements in the ways of doing things, is but one aspect of social evolution. This is so because technological development, as it is conceived and carried out by human minds, depends crucially on how effectively a society can exploit the fruits of the phylogenetic learning of its members. If a society's technological development is inseparably linked with its cultural evolution, it is because cultural evolution is nothing but the process of phylogenetic learning of a society. And to the extent that there is an inseparable linkage between technology and culture, it becomes meaningful to talk about the "character" of social inventiveness in technological development.

In the early days of the evolution of capitalism, it may be recalled, technological development came mostly from the wits and inventiveness of isolated entrepreneurs. Technological development in those days involved, as Schumpeter so aptly characterized it, creative destruction—the introduction of new ways of doing things to supplant old ways.[14] It may also be recalled that the development of capitalism paralleled with the emergence of democracy to supplant autocracy of one form or another on the political scene. This is, of course, no coincidence, for democracy provides a most effective channel through which individual inventiveness of isolated entrepreneurs is translated into social inventiveness.

Creative destruction, however, does not quite capture the character of social inventiveness in technological development in Meiji Japan where initiative and guidance came from the government. To be sure, certain traditional ways of doing things were supplanted by new ways as superior foreign technologies were introduced. But the old, traditional ways of doing things, especially spiritual values, never completely disappear from Japanese culture because it reflects an animistic world view. To the extent that the new ways of doing things have been introduced to supplement, rather than to supplant, old ways, the character of social inventiveness in technological development since 1868 can be better described as "creative assimilation."

Creative assimilation combines both absorption and adaptation. The process begins with the absorption of existing technologies, whether inherited or imported. In the case of imported technologies, absorption is tantamount to imitation. However, Japan is no exception when it comes to imitating imported technologies. Throughout the Middle Ages European countries constantly imitated foreign technologies as they imported, for example, the windmill from Persia, the compass from the Arabs, and gunpowder from China. As was the case with these European countries, the process of creative assimilation in Japan also involved the adapation of imported technologies to suit domestic conditions. This aspect of creative assimilation in Japan, however, involves such a profound transformation of the original technology that it can only be characterized as the process of "Japanization." And B. H. Chamberlain, who went to Japan in 1873 and had a chance to witness the transformation of society in Meiji Japan, reports in his *Things Japanese* how he was struck by this process of transformation which we are here calling "Japanization:"

> The Japanese genius touches perfection in small things. No other nation ever understood half so well how to make a cup, a tray, even a kettle, a thing of beauty, how to transform a little knob of ivory into a microcosm of quaint humour, how to express a fugitive thought in half a dozen dashes of the pencil.[15]

Transforming whatever material is at hand into a "microcosm of quaint humour" is just what the Japanese excel in; this term is indeed symbolic of the spirit of animism which still influences the thinking of many Japanese today. If a Japanese creation seems to carry a spirit of its own, it is precisely because the worker has put his own spirit into it. How else could we explain the kind of craftsmanship the Japanese workers today display in their cameras, conductors, electronics, and watches? What is produced, although it may be a simple physical object, is nothing but a microcosm of quaint humour, a thing imbued with the spirit of the maker whether produced by an individual craftsman or on an assembly line. To the extent that this spirit originates in an animistic world view, the craftsmanship of Japanese workers today is no different from the craftsmanship of Japanese artisans whom Chamberlain observed a century ago.

All this suggests that the Japanese conception of the role of science in technological development may be different from the Western conception. It should be pointed out that, even in the West, the application of science to industry is a fairly recent phenomenon which began with the Industrial Revolution. In fact, what makes the Industrial Revolution a revolution was a discovery that science makes good business sense—a discovery that can probably be attributed to Matthew Boulton when he brought James Watt to Birmingham in 1775 to join in his venture. Before the Industrial Revolution, science was first and foremost a means of understanding the world. It is also worth pointing out that the Scientific Revolution which preceded the Industrial Revolution had introduced to Western man the mechanistic outlook of the world as represented by Newtonian mechanics. It is such a mechanistic conception of the world that also played a key role in the Industrial Revolution, for the very success of the Industrial Revolution, as measured by productivity gain, came from the mechanization of production processes.

To the Japanese in Tokugawa Japan, the mechanistic outlook of the world was not only alien but totally repugnant. For the Japanese, understanding nature did not mean the discovery of the laws of nature, but the discovery of the relationship between the outer phenomena and the inner being, the linkage between the "spirit" of the universe and that of the individual man. Thus, the idea of isolating the object of observation from the observing subject, which was a key to success in classical Western science, was alien to the Japanese. This explains why Japan did not experience a scientific revolution in the sixteenth and seventeenth centuries.[16] This, of course, does not mean that the Japanese lacked the scientific mind. In fact, Japan's success in industrialization since 1868 is testimony to the capacity of the Japanese to absorb the mechanistic outlook of the world which was inherent in the imported technologies.

Needless to say, the Japanese do not stop at absorbing what is imported from abroad. They go on to adapt imported technologies to suit domestic conditions. Indeed, it is here at the stage of adaptation that what is foreign is "Japanized" into a "microcosm of quaint humour" and, therefore, technological development becomes inseparably linked with cultural evolution. It is also here at the stage of adaptation that the relation between technology and man is reflected in the character of social inventiveness in that cultural evolution is very much dictated by the dominant world view. To the extent that an animistic world view has guided the Japanese to this day, it is not difficult to see why the introduction of a new technology does not by itself pose a threat to workers. Once a new technology is well understood, it can be adapted to complement the efforts of the workers to raise productivity. A technological development that does not disturb the relation between technology and man is, in fact, what creative assimilation is all about. The character of Japanese inventiveness in technological development is thus deeply rooted in the spirit of Japanese culture.

Conclusion

Why does the Japanese carpenter pull, instead of pushing, his plane and saw? To this question we can still give only speculative answers. But to the question, "What does this simple act of the Japanese carpenter symbolize?" we can now give fairly satisfactory answers. It symbolizes the act of absorption which the Japanese have exploited to their advantage in the process of their phylogenetic learning as a nation. It also symbolizes the act of adaptation which the Japanese have employed to transform imported technologies into those forms suited to their domestic conditions. More importantly, it symbolizes the intimate relationship that the Japanese have tried to maintain between technology and man, for the act of pulling reminds us of the human proportions of the technologies employed by the Japanese. Thus in this simple act of the Japanese carpenter who pulls his plane and saw towards him, we find a symbolic expression of the character of Japanese inventiveness in technological development as well as of the animistic world view inherent in Japanese culture.

It is by these acts of absorption and adaptation that the Japanese have been able to transform whatever material that has been available to them into a microcosm of balance, beauty, and precision. Moreover, absorption and adaptation, as the fundamental traits of the Japanese mind, are deeply ingrained in the spirit of Japanese culture which has evolved out of an animistic world view. This animistic world view, in that it treats man as an integral part of nature, serves as a restraining force by preventing their technologies from growing out of human proportions. Needless to say, the physical size of a technology cannot always be kept within human proportions; its spirit, however, can be incorporated into the intimate relationship which the Japanese strive to maintain between technology and man.

To the extent that Japan has been successful in economic and technological development with her characteristic inventiveness, it would be natural to ask whether the Japanese experience can serve as a model for other countries. For a country that is trying to take off on a path to economic development, the Japanese experience illustrates how absorption and adaptation of imported technologies can be exploited to its advantage. But it must be recognized that Japan owes her success to the uniqueness of a culture which has been able to maintain with the blessing of the racial and mental homogeneity of her people. This homogeneity has provided the Japanese an environment most favorable for communication and diffusion of information among themselves.

The homogeneity of society, on the other hand, turns against those

Japanese who for one reason or another become isolated from the rest. As the uniformity of purpose and action is forced on all members in such a society, those who find it difficult to accept the demands for uniformity will feel all the more alienated from the society. The Japanese have, of course, found ways of diverting these feelings of alienation. Just as the disgruntled sons of samurais and merchants in Tokugawa Japan sought relief from their daily toil in the gay quarters, the disillusioned Japanese workers today try to relieve tensions of their daily work in the daily rounds of drinking and merrymaking in the back alleys of Tokyo and other towns across the land. The pathology of the Japanese mind is thus caused by the demands for uniformity that the society imposes on its members; a typical Japanese will spend his life in a prolonged period of moratorium in a society in which the cultivation of his individuality is constantly frustrated.[17] Thus for a country that cherishes individuality and, therefore, depends heavily on individual inventiveness in technological development, the Japanese experience illustrates how difficult it is to translate the wits and inventiveness of isolated individuals into social inventiveness without the blessing of the homogeneity of its members.

To sum up, what is social inventiveness in technological development? It is, we believe, that quality of a society which makes it possible to exploit the fruits of the phylogenetic learning of its members. If so, the need for social inventiveness is stronger today than ever before considering the kind of world we live in—the world which is so dangerously divided politically and ideologically yet so intricately interconnected economically and ecologically. In such a world, technological development for the sake of technological development is no longer at stake; we do, in fact, possess the kind of technology that could erase all that we have accumulated during the last few million years of our existence as a species. Whatever destiny our species may be headed for, we must try to develop the ways of doing things such that the fruits of our phylogenetic learning as a species can be exploited for the betterment of our lot in this divided yet interdependent world. Whether we will succeed in this will be the ultimate test of social inventiveness that confronts all of us as a species.

Notes

1. L. Hearn, *Japan: An Attempt at Interpretation* (New York: Grosset & Dunlap, 1904), p. 10.
2. L. Hearn, ibid., p. 11.
3. See, for example, D. W. Jorgenson and M. Nishimizu, "U.S. and Japanese Economic

Growth, 1952–1973," Discussion Paper Number 566, Harvard Institute of Economic Research, Harvard University, Cambridge, Massachusetts, 1977.

4. L. Hearn, *Japan*, p. 11.

5. The stagnation thesis for Tokugawa Japan is especially prevalent among Japanese historians in the Marxist camp. There are also many Western historians who subscribe to the same thesis. See, for example, G. Sansom, *A History of Japan, 1615–1867* (Stanford: Stanford University Press, 1963), and M. Hane, *Japan: A Historical Survey* (New York: Charles Scribner's Sons, 1972).

6. Not all studies agree on the magnitude of population increase in Tokugawa Japan. The problem here, of course, is the lack of reliable data prior to 1721. For a critical appraisal of demographic change in Tokugawa Japan, see Chapter 3, Aggregate Demographic Data: An Assessment, in Hanley and Yamamura (1977).

7. This quotation is from R. P. Dore, *Education in Tokugawa Japan* (1965), p. 16.

8. The exact figure is difficult to obtain because the concept of literacy rate was absent. This figure is based on the ratio of pupils registered as attending primary school.

9. For a discussion of this subject for European countries, see C. M. Cipolla, *Literacy and Development in the West* (Harmondsworth: Penguin, 1969).

10. See A. Gershenkron, *Economic Backwardness in Historical Perspective* (Cambridge: Harvard University Press, 1962).

11. The letter of St. Francis Xavier to Goa as quoted in J. Brodrick, *Saint Francis Xavier* (London: Burns Oates, 1952), p. 362.

12. The stories of Daikokuya Mitsudayū and other Japanese who came back from foreign countries against the edict are reported in K. Tsurumi, *Kōkishin to Nihonjin* (Curiosity and the Japanese), Tokyo: Kōdansha, 1972, Chapter 6.

13. See C. G. Jung, *The Archetypes of the Collective Unconscious* (Princeton: Princeton University Press, 1968).

14. See J. A. Schumpeter, *The Theory of Economic Development* (Cambridge: Harvard University Press, 1934).

15. B. H. Chamberlain, *Things Japanese*, 5th ed. (London: Murray, 1932), p. 34.

16. For an alternative explanation, see J. Bartholomew, "Why Was There No Scientific Revolution in Tokugawa Japan?" *Japanese Studies in the History of Science 15*, 1976.

17. For the concept of "moratorium," see E. Erikson, *Identity: Youth and Crisis* (New York: W. W. Norton, 1968).

References

R. N. Bellah, *Tokugawa Religion* (Boston: Beacon Press, 1957).

R. P. Dore, *Education in Tokugawa Japan* (Berkeley: University of California Press, 1965).

S. B. Hanley, and K. Yamamura, *Economic and Demographic Change in Preindustrial Japan, 1600–1868* (Princeton: Princeton University Press, 1977).

H. Kato, "The Significance of the Period of National Seclusion Reconsidered," *Journal of Japanese Studies 7*, 1981.

T. Koizumi, "The Ways and Means of the Gods: An Analysis of Japanese Religion," *Journal of Cultural Economics 3*, 1979.

W. W. Lockwood, *The Economic Development of Japan: Growth and Structural Change, 1868–1938* (Princeton: Princeton University Press, 1954).

L. Mumford, "Authoritarian and Democratic Technics," *Technology and Culture 5*, 1964.

G. Needham, *The Grand Titration: Science and Society in East and West,* (London: George Allen and Unwin, 1969).

H. Patrick, and H. Rosovsky, *Asia's New Giant: How the Japanese Economy Works* (Washington: Brookings, 1976).

D. H. Shively, ed., *Tradition and Modernization in Japanese Culture,* (Princeton: Princeton University Press, 1971).

chapter eleven

HUMAN FACTORS AFFECTING INNOVATION AND PRODUCTIVITY

Michael Maccoby

W e are all aware that American industry must perform in a highly competitive international economy. This requires higher levels of productivity and product quality, calling for both new production technology, new styles of management and social inventiveness.

A clear example is the auto industry. The Japanese understood the American market better than did GM, Ford, and Chrysler. When the price is right, quality and durability sell, especially during inflationary times. This requires workers who care and have time to fix mistakes and management that invites them to help solve problems and assures them that a cooperative attitude will not be exploited.

We have known for a generation that the management of technological innovation requires teamwork among scientists, engineers, and marketing.[1] Yet, many companies still have not learned the lesson. According to the *Wall Street Journal* (September 3, 1981), "More and more companies are climbing aboard the high-technology bandwagon, but most of them don't have the slightest idea how to manage their technology efforts effectively."

We are now learning that teamwork is also required for productivity in factories and offices. Studies based on opinion polls, in-depth interviews, and experiments in new forms of management all indicate a significant shift in the values and attitudes of workers.[2] The new breed workers demand to be treated with respect as thinking, responsible people rather than as either children or machine parts. They dislike taking orders without good reason and no longer are controlled by an awe of authority. They want to learn at work and have a say in how it is carried out. Otherwise, they become resentful and cynical, game-players who figure out ways of beating the system. For such workers, hierarchical, policing-style management causes sabotage, costly absenteeism, and a negative attitude to the business. The Japanese success at participative management and the GM-UAW Quality of Working Life (QWL) program have dramatized the fact that properly organized, workers today can manage themselves, raising the level of performance and reducing the costs of administrative overhead and waste as they also find work more satisfying.

The GM success at their Tarrytown assembly plant proved the value of a QWL program. Starting at the bottom in terms of productivity and quality measures, Tarrytown moved near the top of GM's plant ratings.[3]

If adequately trained and informed, factory and office workers contribute to a continual process of innovation. Small improvements and cumulative savings add up and can be just as important as more dramatic innovations.

Particularly in jobs that call for brainwork, management must establish an environment of trust and support, in which employees feel free to be flexible and innovative. This must be achieved without losing control over the organization, calling for what Harvey Brooks terms "managerial innovation."[4]

In 1980, the Department of Commerce organized a meeting, "The Frontiers of Management," bringing together companies that were developing a management capable of achieving an effective and productive organization.

Two types of companies have been at the frontier of managerial innovation in the United States. One type is in many ways similar to Japanese paternalistic firms. These are companies that guarantee job security and manage according to a philosophy which emphasizes respect for the individual and encourages continual learning. Many include profit-sharing. Hewlett-Packard, IBM, and Procter & Gamble are examples. In these companies, employees cooperate to develop and implement productive new technology, because they do not fear they will lose their jobs. They are willing to experiment with new organization and job classifications because they trust management will not abuse their cooperative attitude.

The second type of innovative management is to be found in those unionized companies that are able to work cooperatively with a strong and progressive union. General Motors, the UAW, AT&T, and the CWA are examples. By working in a "limited partnership" with the union, management gains cooperation and flexibility. Because their rights are protected and attention is paid to their concerns, workers are willing to participate in cutting costs and improving quality. Unions are flexible about job classifications and the introduction of new technology, and they put pressure on employees who take advantage of loose controls.

As American business reaches for new technology to improve productivity, it must re-examine its approach not only to industrial relations, but also to design and engineering. Traditional production technology like the assembly line, built to maximize control and the interchangeability of people, no longer proves productive when compared with new socio-technical designs such as those that have been built on the principle of increased teamwork and individual responsibility at Volvo in Sweden and GM in the United States.

The Volvo plant at Kalmar was the first to develop a team approach to auto assembly, using carriers rather than an assembly line to move the car from station to station. Teams of workers can vary their cycle-time and pace, with buffer areas that allow them to time their own breaks.

This socio-technical approach has been adapted to other Volvo factories. In each case, new technology such as the carriers and improved material handling systems are combined with managerial innovation. At one engine assembly plant, teams of workers have learned to take over managerial functions such as personnel, maintenance, planning, quality, etc. In a factory of 100 assemblers there are now only two engineering managers and a head plant manager who believes he has no further function and has suggested that his job be abolished.

New forms of socio-technical design become more feasible and necessary as computers and microprocessors allow decentralization and flexibility in designing away some of the worst jobs. In the future, as more routine jobs can be performed by robots, new jobs will demand brain work, intellectual skills, requiring greater autonomy and responsibility in offices as well as factories.

Technology will not fulfill its promise of productivity without new attitudes and practices by management and union leadership. Professor Richard Walton at the Harvard Business School has presented research showing that unless employees participate in the design and implementation of new office technology, costly resistance depresses performance.[5] Despite evidence to the contrary, some business and government agencies expect new technology to work efficiently according to the old hierarchical, controlling managerial principles. At the Department of State, Ruth Schimel has shown that participation allows offices to make good use of new office technology, while traditional autocratic management can result in break-downs and confusion.[6]

It is significant that when he reviewed the findings on the Three Mile Island near disaster, John G. Kemeny concluded that the failure was caused by human and managerial, not technical systems.[7] There was a management attitude that nothing could go wrong. In a nuclear power plant, as in many large and delicate continuous processing facilities, workers who run the machines should have the training to handle emergencies and authority to make quick decisions. The hierarchical chain of authority is not only cumbersome and costly, but dangerous.

In government as well as business, managerial innovation pays off. City workers in Springfield and Columbus, Ohio achieved a high level of cooperation and innovation in a QWL program developed by Ohio State with the participation of AFSCME. With guarantees that they would not be laid off because of improved productivity, refuse collectors in Springfield organized themselves into teams with CB radios so that those who finished their routes early could help others. In Washington, I helped to establish a work improvement program at the office of publications in the Department of Commerce with the cooperation of the Federation of Federal Employees. Workers met to identify and solve production problems. The program resulted in significant productivity gains and technical innovations suggested by workers. Grievances practically disappeared.[8]

Given that new approaches to management pay off, why don't more companies and government agencies adopt them?

One answer is that change is difficult, especially for managers with inflexible, rigidly controlling personality traits. Even those who are more flexible must reorient themselves and question many of their ideas about people and much of their training. It is easier to believe rumors that new programs have failed than to change one's deep rooted attitudes.

The work of Rensis Likert has been especially valuable in providing methods for managers to examine critically traditionally directive methods and style of management and to try out more consultative and participative approaches.[9]

Change requires time, education, trial and error. Edwin Land of Polaroid has pointed out that organizations do not treat social research and development with the same scientific detachment and persistence shown with innovations in hardware. No one expects the first attempt at building complex technology, such as a new computer system, to work without adjustments and changes, trial and error. The memory system and central processing unit may be redesigned a number of times to fit each other. Nor does anyone expect even the best computer system to be the ultimate model. Yet, when new experiments at managerial innovation fail or are only partially successful, someone will always say, ''See, it doesn't work, because people can't accept responsibility.''

They fail to ask whether proper conditions for socio-technical innovation were met, whether employees had the proper training, whether everyone understood the goals and values directing the change, and whether all levels of authority were included. Studies of successful managerial innovation indicate that success depends on factors such as these. Indeed, the history of attempts to improve industrial work should be studied, starting with the methods of Frederick W. Taylor and scientific management to Elton Mayo, the human relations approach, and more recent socio-technical developments. By analyzing the successes and failures of these managerial innovations, we learn what it takes to create a higher level of employee involvement and where attempts tend to flounder.

Many companies, particularly in the United States, find it difficult to establish the necessary relationships with unions. In unionized companies, union cooperation is essential to develop participative management or to institute Japanese-style managerial innovations such as quality circles, in which groups of workers meet to identify and solve production problems. When unions are left out, they see these new approaches as threats to their traditional role, and typically, they undermine them.

Another cause of resistance is lack of information. We need good models of managerial and socio-technical innovation with case histories that can be used for teaching. Also useful are studies that demonstrate economic, human, and social dangers of new technology that is poorly managed.

However, it is not easy to obtain adequate and useful data about model programs. There are some companies, such as Procter & Gamble, that consider socio-technical innovation to be proprietary information.

In the case of Volvo, there have been rumors that the socio-technical innovations begun at the Kalmar plant have been a failure and abandoned. In fact, this highly innovative approach is 20 percent more profitable than traditional assembly facilities at Volvo, and a higher level of cooperation between the local union and management has been achieved.[10]

What should be the role of government in furthering managerial innovation? In the past, governments have stimulated and supported pilot programs. In Norway, the Industrial Democracy Project had tripartite support from government, unions, and employers. In the 1960s, that project studied both the economic and human effects of different forms of participation in Germany and Yugoslavia and helped to develop new models in Norwegian factories. In 1973, Einar Thorsrud, director of the

Industrial Democracy Project advised those of us who were working with Harman Industries and the UAW at Bolivar, Tennessee, on an American model of union-management cooperation to improve work. The Bolivar Project was partially supported by the National Commission on Productivity, which provided neutrality required by the union. Government funding also allowed researchers to study in depth human, social, and economic factors affecting the project. One finding was that different workers were satisfied and dissatisfied by different aspects of work. While some workers sought more demanding work with possibilities for career development, others preferred less complex jobs that allowed sociability at work. While some workers wanted to take rewards for improved productivity in higher wages, others preferred time off to run small farms, work as homemakers, or go hunting and fishing. Taking account of differences in values and goals avoids the error of basing change strategies on an ideological view of motivation, such as the idea that all workers want more challenging work.[11]

Government can play a needed role in supporting innovative new approaches to organizing technology and work and understanding them. But to be useful, such research must itself be innovative. In my experience, social scientists can make themselves a nuisance by their compulsion to measure everything, including the unmeasurable. Some variables can be measured usefully; others require more anthropological study and analysis in the form of case histories that explore culture, human motivation, values and industrial relations as well as technical and economic factors.

Einar Thorsrud has written:

> the segmentation of scientific disciplines and professions is a serious constraint in technological and social changes. It can be overcome when members of research teams coming from different disciplines are made *jointly responsible,* together with employees in enterprises, for the evaluation and utilization of results. The separation of schools from the world of work offers a similar problem. The integration of maritime education into the shipping change program has now been achieved by involving the schools directly in the early phases of new shipping projects. Large investments in purely disciplinary institutes are not likely to be effective. Such investments tend to help already closed institutions and professions to maintain their privileged roles and their lack of involvement in social change. In real life change projects the distinction between basic or fundamental research and applied or action oriented research is artificial. Good applied or action oriented research must include phases of improvements in theory and method, and is in this way dependent on basic research.
>
> Formulation of research policy and allocation of research resources cannot be based primarily on decisions in representative committees. Steering

committees can play a useful role in development projects. A major force in science policy making can come from scientific institutions themselves when they share their research tasks with those directly affected. . . . Such a democratization of the research and development process is particularly important when choice of new technology and choice of new organizational form is involved.[12]

The ideal role for the social scientist in socio-technical innovation is as a co-learner who helps participants develop their own approach to study and shares his own observations. By describing models that have worked elsewhere and criticizing the participants' self study, the researcher also becomes a teacher who helps create a capacity for self-evaluation which becomes feedback useful to planning.

Recently, I attended a small meeting of the National Research Council chaired by Irving Bluestone to discuss methods of studying union-management programs to improve the quality of working life. The group included representatives from unions, management, and academia. All agreed that evaluative study would be useful to gain knowledge about what works and to encourage new programs. But the group concluded that such research is ideally part of the QWL process itself and should be designed and implemented by the participants. The role of the National Research Council could be to design a basic evaluation instrument, including key variables and different methods that have been employed to measure them. This would be made available to companies and unions with the request that they make the results available to the National Research Council.

Among the human factors affecting innovation and productivity, one of the most important is the quality of leadership. Managerial and socio-technical innovation has required leaders. They have had to sell skeptical people a vision of a new approach to work. In many cases, they have been motivated not only economically, but also by a concern for people. They have felt a deep revulsion against the wastage of human life in dehumanizing jobs.[13] In leading innovative projects, they have both articulated new values and designed practical strategies to change organizations and create socio-technical systems. Pehr Gyllenhammar, the chief executive of VOLVO, instructed a design team including engineers, a sociologist, and worker representatives to create a new factory, taking into account human and social as well as economic and technical criteria. In so doing, he faced skepticism of traditional engineering management. Irving Bluestone, vice president of the UAW, described QWL as allowing the union to expand its traditional collective bargaining role and better serve its members. At the start, he was almost alone on the union's executive board in believing such an approach would work. Most union leaders distrusted any deviation from traditional adversarial bargaining, lacked competencies required by the new relationship, and believed companies would use new programs to manipulate workers.

Some managerial innovations produce technology, social forms, and training that institutionalize the new system. It is harder to institutionalize leadership, although the concept can be usefully demystified. Cooperation does not require paternalistic or charismatic leaders. With adequate training, a group can exercise many leadership functions, such as planning, budgeting, quality control, and evaluation, but there will always be need for individual leaders who articulate and defend the values that are essential for trust and participation.

Notes

1. T. Burns and G. M. Stalker, *The Management of Innovation* (London: Tavistock Publications, 1961), and Paul Lawrence and Jay W. Lorsch, *Organization and Environment: Managing Differentiation and Integration* (Boston: Division of Research, Harvard Business School, 1967).
2. Michael Maccoby and Katherine A. Terzi, "What Happened to the Work Ethic?" Joint Economic Committee of Congress, 1979.
3. Robert Guest, "Quality of Work Life—Learning from Tarrytown," *Harvard Business Review,* July-August 1979, p. 76–87.
4. See Chapter 1 of this book.
5. Richard Walton and Wendy Mela, *Explorations,* Boston, Harvard Business School, no. 19, June 1980.
6. Testimony of Ruth Schimel. Subcommittee on Science, Research and Technology, September 1981.
7. John G. Kemeny, "Saving American Democracy: The Lessons of Three Mile Island," *Technology Review,* June/July 1980, p. 65–75.
8. Michael Maccoby and Robert and Margaret Molinary Duckles, *Bringing Out the Best,* Discussion Paper Series, JFK School of Government, Harvard University, No. 91D, June 1980.
9. William F. Dowling, "At General Motors: System 4 Builds Performance and Profits," *Organizational Dynamics,* Winter 1975, p. 23–38.
10. Testimony of Berth Jonsson. Subcommittee on Science, Research and Technology, House Committee on Science and Technology, September 1981.
11. Michael Maccoby, "Changing Work, The Bolivar Project," *Working Papers,* Summer 1975, p. 43–55.
12. Testimony of Einar Thorsrud. Subcommittee on Science, Research and Technology, House Committee on Science and Technology, September 1981.
13. Michael Maccoby, *The Leader: A New Face for American Management* (New York: Simon and Schuster, 1981).

chapter twelve

THE JAMESTOWN EXPERIENCE: A CASE STUDY IN LABOR-MANAGEMENT COOPERATION

Stan Lundine

The community or areawide labor-management committee is a significant social invention which has considerable potential for improving productivity performance at the micro-economic level. The "spin off" or ancillary effects of such an areawide labor-management program can create entirely new work systems and job training concepts. The prototype of such a labor-management committee concentrating on improving the quality of work life, rather than simply existing for the purpose of mediating labor disputes, was developed in Jamestown, New York in the early 1970s. My participation in commencing this venture began in an unusual way.

On a dreary day in late November 1971, I encountered an old Swedish resident of Jamestown who asked me why I looked so terrible, "Well", I replied, "no sooner do I find out that Art Metal has gone bankrupt throwing 900 people out of work, than I learn that Crescent Tool will probably leave this area for the South taking with them another 700 or 800 jobs."

My older friend pointed down into the industrial valley along the Chadakoin River where several old factory buildings were clearly visible and said, "See that old building down there. I remember when the worsted mill closed and everyone said 'Jamestown is dead.' Then, in came the wood furniture factories and the same people found work at higher wages. Later, almost all the furniture factories moved to the South and everyone said 'Jamestown is dead.' Then, along came the metal manufacturing companies and everyone found work at higher wages and with better conditions. Now, Mayor, all you have to do is figure out what should come along after the metal manufacturers."

Coincidentally, it was later that same afternoon that a local labor lawyer stopped into my office to discuss the city's future. He said that he felt we needed a real dialogue between union leaders and business executives if Jamestown was going to reverse its gradual economic decline which had now accelerated at a dangerous pace. "You've just been re-elected by an overwhelming majority. You have strong support from both the business community and the major labor leaders. You can take the leadership and call these people together in a nonadversarial situation and get them to work together."

These two people, very different in their perspectives, both had given me a glimpse of a revitalization possibility and the essence of an effective industrial strategy. Their observations prompted me to explore the possibility of joint labor-business-government action to address our community's critical economic problem.

Jamestown is a small city of about 35,000 people and serves as the employment center for an area approximately three times that size. It is heavily dependent on manufacturing employment, about 80 percent of which is unionized. Located in the far southwestern portion of New York State, it is south of Buffalo, north of Pittsburgh, and east of Cleveland. Yet, it is an independent economic entity and not really a satellite of any metropolitan area.

The crisis in the local economy in the early 1970s was preceded by two decades of a gradual erosion of manufacturing jobs. Obsolescent factory facilities, outdated management practices, high taxes, and other burdens of doing business in New York, as well as labor-management strife, had contributed to the economic decline. There can be no doubt that labor-management relations were poor, but the reputation Jamestown had acquired as a bad labor town probably went even beyond the bitter facts. Nevertheless, we had experienced about 2-1/2 times as many lost days due to strikes during the 1960s as the average in New York State. Business and union leaders rarely exchanged any ideas outside of the collective bargaining process.

The residue of bitterness was so intense that when I called the first meeting of the manufacturing executives and labor leaders in City Hall in January of 1972, I decided to have them meet separately. The result of those sessions was, however, a resolution to meet together and discuss the possibility of joint labor-management action on some of the economic problems of the community.

The first meeting of what was to become the Jamestown Area Labor-Management Committee (JALMC) occurred in February, 1972. At first, these men who were unaccustomed to exchanging ideas except as adversaries at the bargaining table were somewhat awkward and restrained. We really began making progress when a union leader accused the manufacturers association of systematically trying to keep all new business out of the area in order to keep wage rates low. Several business executives angrily denied this charge and began to enumerate their reasons why business activity was on the decline. At the end of this spirited dialogue, it was resolved that the labor-management committee would work with the Mayor and county industrial development agencies to not only revitalize existing businesses but attempt to attract new industry to the area.

The fundamental objectives of the labor-mangement committee were idenitifed as: (1) industrial development, (2) improving labor-management relations, (3) manpower development, and (4) productivity improvement. It was decided that we should first concentrate on improving productivity in existing industry. Only by doing so could we expect to achieve industrial development and create a base for improving labor relations.

Naturally, the productivity concept did not come easily to some of the union leaders on the committee. There were several clear understandings regarding the productivity objective that were worked out at the very outset of the Jamestown program.

First, we agreed that productivity would be broadly defined and would not simply involve a speed up of worker effort. Such objectives as reducing absenteeism, improving product quality, and eliminating waste would be recognized as productivity advances.

It was also understood that labor and management would work together on the development of programs and would share equitably in the results of any improvements. Without any up-front negotiation, it was agreed that the financial gains achieved as a result of the joint productivity efforts would be shared on a roughly equal basis between labor and industry.

The third assurance regarding the productivity goal was probably the most important. It was agreed that no jobs would be lost in an industry as a result of joint efforts to improve productivity in that enterprise. While it was clearly understood that some jobs might be eliminated in the process being undertaken because of improved effectiveness, management accepted the concept that these workers would be transferred to other jobs within that enterprise.

We had some excellent speakers at some of the early dinner meetings, including Hobart Rowan, the economics editor of the Washington Post, and Dean Robert McKersie of the Industrial Labor Relations School at Cornell University. These were opportunities for the entire labor-management committee in Jamestown to become acquainted with each other and to be exposed to interesting new ideas on cooperation between unions and business after having defined mutual objectives.

During the course of these dinner discussions, it was decided that the labor-management committee would devote its attention to areas of cooperation that might be achieved apart from the collective bargaining process. It was definitely decided that we were not critically needed to mediate labor disputes. Moreover, we recognized that the areawide committee had a much better chance of success if it did not become involved in collective bargaining. The improvement of labor relations would come as a natural consequence of concentrating on those areas where business and labor had common goals. We recognized that in some cases collective bargaining agreements might have to be modified in order to carry out some of those jointly designed projects, but clearly specified that the labor-management committee was not intended to settle strikes.

To achieve definite results, we quickly understood that it would be important to form labor-management committees in individual plants where both the union and the management were receptive to cooperation. Typically, these in-plant committees would consist of six or seven management employees ranging from foreman to the chief executive, and the union would have a like number, some of whom would be union officers while others simply volunteered to participate as labor members. The areawide committee was essential in establishing a community atmosphere of cooperation and in setting forth general goals. But the committee recognized that serious efforts to improve productivity would have to take place in the work places themselves.

Recognizing that it would not be possible to provide assistance to these in-plant labor-management committees with amateurs, we decided to apply for federal assistance to fund consultants who might help facilitate the productivity goal. For many months, we were frustrated by bureaucratic barriers in our efforts to obtain even modest support for what we were convinced was an innovative industrial revitalization technique.

Oddly enough, this period and the repeated unsuccessful trips to Washington and other prospective funding sources unified the labor and mangement representatives in a way that any early success probably would not have. By the time that a $22,500 grant was obtained from the Economic Development Administration (EDA) early in 1973, the Committee had already been through several trust-building experiences.

In the initial months, several local industries came close to liquidation. The first was Chautauqua Hardware Corporation which was a significant supplier to the furniture industry. Union members were actually persuaded to work for two weeks without pay at Chautauqua Hardware while the company was being reorganized. New professional managers and investors were brought in, arrangements were made for refinancing and for delaying some of the repayment of back taxes and utilities.

The labor-management committee helped to open up a dialogue between the new owners and labor which resulted in an agreement to tie future wage increases directly to productivity improvement. This was followed by the establishment of the first in-plant labor management committee in this company in 1973. With the cooperation of the union members and much more aggressive management, Chautauqua Hardware Corporation exceeded every productivity threshold set up in their original negotiation, and the gains from the improved productivity were shared by labor as well as management throughout the entire period of the contract.

This labor-management in-plant committee also became involved in quality improvement projects and planning departmental physical redesign. The efforts continued over a period of years while employment in the firm almost doubled. By 1978, Chautauqua Hardware, with some help from the city and county industrial development agencies as well as a federal Urban Development Action Grant, constructed the first new foundry facility of its type built in the Eastern United States in 50 years.

By working on this and several other industrial reorganization opportunities afforded by these several failing companies, business and union leaders learned to develop trust and respect. In addition, some of the positive possibilities of labor-management cooperation became clearly apparent. It was also during this period of time that the co-chairmen of the committee, one from management and the other from labor, became the principal directors of an action program. They worked together with the Mayor in designing the communitywide labor-mangement meetings, setting an agenda for regular executive committee sessions and carrying out the industrial revitalization initiatives jointly outlined.

Soon after receipt of the EDA grant, the JALMC decided to retain James McDonnell, a professor at Buffalo State University to serve as its first full-time coordinator. The choice was fortuitous, as McDonnell proved to be a very effective communicator who brought positive methods of labor-management cooperation to the attention of rank and file union members, management employees at all levels, and the community at large.

In addition, we were very fortunate to have the consultation and advice of several people who had expertise in labor relations and provided enormous insight into the possibilities and limitations of the labor-management concept in its early stages. Sam Nalbone, the Ombudsman for the City of Jamestown, for example, had been a business representative for the International Association of Machinists and was a widely respected labor relations professional in the area. He often provided advice tempering some of my more extravagant enthusiasms during this formative period of the committee's development. A commissioner of the Federal Mediation and Conciliation Service in Buffalo was a tireless advocate for labor management cooperation and provided exceptionally keen insight from the vantage point of a neutral third party who had come to know intimately many of the negotiators on behalf of Jamestown labor and business organizations during his years of mediation experience. Various advisors from the Industrial Labor Relations School at Cornell University also provided important strategic advice, along with other university and labor relations specialists.

Some outside advisors (or would-be advisors) proved to be more troublesome than helpful. For example, several social scientists wanted access to every industry to conduct baseline productivity and worker attitude surveys. Local leaders who really made the committee function effectively remarked that the town could go down to economic demise while the experts were gathering their baseline information.

When grant funds were finally available, the JALMC begain a nationwide search for the most effective consultants for the in-plant committees. We finally settled on Eric Trist of the Wharton School at the University of Pennsylvania as the leading organizational development consultant. Dr. Trist and two of his associates, John Eldred and Bob Keidel, provided the first outside assistance in the development if the in-plant labor-management committees and also gave the concept of improving quality of working life its first emphasis in Jamestown's evolving program.

Beginning with his pioneering work with miners in the British coal fields 25 years ago, Dr. Trist has been one of the founders of the Quality of Work Life (QWL) movement. As our experience evolved it became evident that QWL is a participative *process*, by which workers and management solve work problems together. The term encompasses a wide range of activities in which work tasks are reorganized, assembly lines or plant lay-outs are redesigned, and job environments are altered in ways that are more satisfying and meaningful to workers and more effective from the standpoint of management.

The forms of labor-management cooperation devised by the in-plant committees in Jamestown were extremely varied. They ranged from gains sharing as an incentive for increased productivity to reduction of waste and breakage in a glass manufacturing plant, from quality control systems to cooperative design of new products, and from joint effort towards improved bidding procedures to the collaborative design of an entire new factory facility.

An incentive system was devised by the labor-management committee within Falconer Plate Glass Company. Breakage of glass during production had represented a significant cost to the firm, as well as a safety hazard to the workers. Starting in one department, the labor-management committee designed an incentive program to reduce breakage. The plan successfully reduced breakage by over 45 percent during the first year of operation, and soon a companywide gain sharing program was developed. After a decade of progress, Falconer Plate Glass has undertaken a major expansion program in the Jamestown area and cited the labor-management cooperation as a major reason for its new investment.

After a bitter nine week strike in 1976, the Corry-Jamestown Company, employing approximately 500 workers and manufacturing high quality office furniture, asked the Jamestown Area Labor-Management Committee for assistance in developing a cooperative approach to improve plant performance as well as labor conditions. As he had done in many Jamestown plants, then JALMC coordinator, Jim Schmatz, was successful in organizing a labor-management committee addressing several problem-solving aspects of the firm's operation. The first one successfully implemented involved the introduction of IMPRO-SHARE, a productivity gain sharing program similar to the Scanlon plan. For this project, management brought in an outside consultant, Mitchell Fein, the creator of the IMPRO-SHARE program. Management estimated that productivity was improved by 15 percent in 1978 and 22 percent in 1979 as a result of the introduction of this incentive system. This meant worker bonuses of 7-1/2 percent of wages in 1978 and 11 percent in 1979.

Labor-management cooperation at Dahlstrom Manufacturing Company, one of six companies saved from liquidation in JALMC's early years, took a quite different aspect, essentially involving improved quality control systems. Union workers were frustrated at many of their machines by delays caused by the unavailability of quality control inspectors. The labor-management committee successfully worked out a program where workers would do their own inspection, thereby improving productivity and actually reducing the number of substandard machine operations. Dahlstrom has initiated several other labor-management cooperation efforts even involving changes in layout and work organization.

Still another form of in-plant labor management cooperation evolved at American Sterilizer (AMSCO) where cooperative problem-solving led to the design of a new product; in addition, a pilot self-managing work team was established, and a joint approach to product bidding was developed. The labor-management committee at Hopes Windows division of Roblin Industries, a manufacturer of industrial window frames, also concentrated on a joint approach to improved competitive bidding. Before the labor-management venture, Hopes had been successful on only one out of ten bids. Afterwards, the new system raised Hopes' success ratio to five out of ten. This was not simply because workers came to understand the elements of cost and developed new cost savings ideas. This involvement led them to recognize the importance of cost efficiency in protecting jobs. Earlier, in slack periods, workers had slowed down to avoid layoffs.

One of the most exciting outcomes of labor-management in-plant cooperation occurred at the Monofrax division of the Carborundum Corporation near Jamestown where an entirely new plant facility was designed on a cooperative basis by the plant labor-management committee. This committee rejected an initial proposal by consultants to management calling for an investment of over $10.5 million in a new building. Soliciting suggestions from every worker in the plant and actually receiving ideas from more than 80 percent of the employees, local management and union representatives began to redesign the plant. When finally presented to corporate headquarters, this project involved a new facility expenditure of only $5.5 million which became the Carborundum Corporation's first capital improvement priority in 1977. The union as well as local management recognized that this averted another plant shutdown and increased the competitiveness of the firm, thereby improving job security.

The success of the labor-management committee during this period was not limited to in-plant cooperation activities. One of the JALMC's major challenges was provided by the failure of a Jamestown-based small conglomerate called AVM Corporation. Due to a complicated series of financial reverses, AVM found itself insolvent, and in 1975 many of its local divisions were targeted for liquidation. The new chief executive officer, sent in to eliminate all but the most profitable divisions, indicated that he would be willing to consider offers to purchase some of the unwanted AVM operations. At Jamestown Metal Products, 88 out of 120 employees invested in the new company which took over the manufacture of hospital and kitchen metal cabinets. Thus began a successful venture in employee-ownership.

One of the most successful industrial revitalization projects was accomplished at another former AVM division, Jamestown Metal Manufacturing, under the leadership of John Walker. A competent manager, Walker had been previously associated with both Art Metal and AVM. When given the opportunity to own a major share of this firm manufacturing marine equipment and other fabricated metal products, he increased the volume of business so that employment grew from a low point of less than 90 workers to almost 500 today. Walker also served as management co-chairman of JALMC from 1975 until 1979.

In 1977, Dahlstrom Manufacturing Company Inc. with approximately 450 employees also faced the prospect of liquidation. Once again, a local industrial group reorganized this old manufacturing company and obtained the active cooperation of union leaders. Harold Bolton, the new Dahlstrom chief executive, was among the first industrial leaders to accept readily the concept of in-plant labor-management committees, and the understanding that was created through this mechanism was important to redirecting this company towards profitability and success.

The most dramatic industrial development activity of JALMC occurred when, for the first time in at least fifty years, Jamestown attracted a new major company to join its existing manufacturing base in 1974. Cummins Engine Company, among the most progressive manufacturers in the world, decided to move into the one million square foot plant, formerly the site of the defunct Art Metal Corporation. Cummins attributed its decision to the labor-management endeavors in Jamestown and pledged to institute a quality of working life program in its new plant which would give workers greater autonomy and which involved innovative management systems.

At present, Cummins is producing diesel truck engines in its Jamestown plant with about 800 employees and has plans to expand to about 2,000 jobs. It has also provided a great deal of leadership in the area of quality of work life programs in the Jamestown area. In its modern plant, work teams assemble entire engines and regulate their own production process. This plant is not union organized but does work cooperatively with the communitywide goals of JALMC.

Improved manpower development was also an original objective of the overall labor-management committee. This was successfully implemented during the early years of the committee's existence and has been sustained up to the present time. The skills development program was originally designed as a result of Jamestown's declining worker skill base in the wood furniture industry. It was found that the average

age of the skilled furniture worker was over 55 and there were few younger workers prepared to replace retirees.

An innovative skills development program was designed involving several furniture manufacturers, the local community college, and the labor management committee, as well as the United Furniture Workers union and the Manufacturers Association of the Jamestown Area. Teams of teachers included skilled craftsmen as well as college professors. Some of the classes were held in the local factories, while others were conducted at the community college.

This skills development approach led Jamestown Community College to establish a system of brokers who would contact local industry, ascertain the skill training needs that could not be met by the individual enterprise, and then match that need with available programs, often combining several local resources with the conventional approach taken by the Comprehensive Employment Training Program. (CETA).

Educational endeavors sponsored by the labor-management committee in Jamestown were not limited to skills development. It was also decided that it would be important to hold courses in the quality of working life concept at Jamestown Community College, and a training program was commenced in 1978 to develop local talent for the vital third party role in the labor-management process. One alumnus of this program, Richard Walker, joined the JALMC staff under a program sponsored by the Cummins Engine Company.

Several educational opportunities were sponsored by JALMC that provided training for shop foremen and union stewards. Improving the problem-solving capacity of these front line managers and worker representatives not only reduced grievances but improved productivity as well. Particularly after the introduction of QWL programs by the in-plant committees, it was essential that foremen acquire additional skills. These first line supervisors are the most threatened and intimidated by allowing greater worker participation in the design of their jobs and in solving work system problems. Foremen who were used to saying "shut up and do it this way because I say we're going to do it this way" were now being expected to solicit suggestions, foster teamwork, and act as tactical advisors. The redefinition of first line supervisors and the degree of their educational advancement often determined the success of the in-plant projects.

The results of this unique social invention in the form of labor-management cooperation in Jamestown, New York have been indisputably, even dramatically, successful. Not only did unemployment decline from 10.2 percent in March of 1972 to 4.2 percent three years later, but the cooperative endeavors helped our community avoid the worst aspects of the national recession in the 1974–76 period. Prior to this time, it had been said that "when the nation's economy catches a cold, Jamestown gets pneumonia." However, in the entire period following the increase in manufacturing employment in 1975, Jamestown's unemployment has remained well below state and national averages. Old firms have been revitalized and new ones attracted to the area. As a result of joint efforts to improve productivity and the quality of working life, there were new jobs and more interesting approaches to work at older ones.

The labor relations climate in Jamestown also improved markedly. During the first three years of the labor-management program, only eight working days were lost due to strikes. For the entire decade of experience, the city's record of work stoppages was substantially better than the state average. Grievances and absenteeism were also significantly reduced. The reason was obvious: trust was being established between labor and management. While the JALMC did not intervene in collective bargaining, the relationships established at the annual steak fry and other informal meetings of the areawide committee were resulting in new avenues of communication and confidence. More than once the idea of entering into an early contract negotiation was broached at one of these informal sessions.

The collaboration between labor and management in the manufacturing sector also spread to encourage other forms of community participation. Jamestown was designated as an All-America City in 1973, partially as a result of its labor-management activities. Soon the school system, the community hospital and local utilities began to experiment with labor-management cooperation.

Most importantly, the intangible factor of community attitude was dramatically reversed primarily due to the activities of the labor-management committee. Outside attention came from such sources as *Newsweek, Business Week,* and the *Wall Street Journal.* They not only gave credibility to the labor-management program, but bolstered Jamestown's confidence in its ability to solve its own economic and social problems.

An attitude of optimism and a spirit of joint problem-solving permeated far beyond the workplace. People began to demand improved quality of life outside of work as well as an improved quality of working life. The most dramatic physical evidence of this spirit of revitalization occurred in the downtown area where the city's urban renewal efforts led to the construction of a major new hotel and attractive pedestrian malls which provided an atmosphere for retail and commercial success in the central business district. More importantly, people began to participate to a greater degree in community affairs. For example, neighborhoods banded together to rehabilitate older homes and provide social, educational, and cultural opportunities for poor people. Here was an old factory town in America's industrial scrap heap which felt a new vitality and a new commitment to work together to solve remaining community problems.

The implications of the Jamestown experience seem clear. It serves as a prototype for an areawide labor-management committee which encourages in-plant problem solving activities. Similar efforts have been undertaken in Evansville, Indiana; Cumberland, Maryland; Buffalo, New York, and several other communities.

The areawide labor-management council provides a very effective instrument to allow for continuing cooperation, even when the in-plant programs experience their inevitable peaks and valleys of activity. Sometimes the in-plant activity is disrupted due to a labor dispute or some totally outside factor such as the sale of the business or a change in management. At these times, this joint committee can continue to participate in the areawide activities rather than dissolve its ongoing function as a problem-solving instrument within the individual enterprise.

It is important to recognize that government's role in such an areawide committee should be as a catalyst and moderator, not the final arbiter of the fundamental objectives of the endeavor. Some of the success in Jamestown was due to the willingness to keep the labor-management council out of partisan politics. Government became a limited partner in the process.

The Jamestown program also demonstrated the special value of a third party, not only in defining the mutual goals of an areawide committee but in serving as a consultant to the in-plant committees as well. As Mayor, I was often called upon to nudge one side or another when communication broke down or old suspicions returned. In fact, even today after ten years of experience, the Mayor's role as a moderator and facilitator is crucial to JALMC's program. It was very clear from the beginning that the unions would only regard labor management cooperation as a truly joint endeavor when the initial funding came from third party sources. If business funding dominated, they too often felt they had been tricked by management devices that ended up ignoring worker desires.

Growing out of my own experience in Jamestown, I sponsored the Labor Management Cooperation Act which was passed at the end of the Congressional session in 1978 with the enormous assistance of Senator Jacob Javits. This program, which is now funded at the modest federal allocation of $1 million for the first year in 1981, provides small grants for community, industry or even plant-level labor management projects that need outside funding to begin a process of mutual collaboration. It was very gratifying that over eighty applications were submitted to the Federal Mediation Service which administers this program when only abut nine or ten could be approved this year. Inquiries were received from more than 400 labor-management committees which have sprung up across America. Their interest in this federal program suggests a recognition that there is a crucial role for the third party as a catalyst to allow labor and management to achieve their shared goals.

The second major implication of the Jamestown experience is that productivity improvement is *the* basic element in effective economic revitalization efforts. Definitions and understanding of productivity must be very broad, and there must be a healthy respect for the attitudes of labor as well as management. However, only when output increases can competitiveness be enhanced and economic revitalization really achieved. In this connection, it should be recognized that the human factor in achieving increased productivity is at least as important as capital investment, regulation, and technological innovation.

In Jamestown, we were fortunate to have the leadership of an absolutely outstanding and courageous labor co-chairman who was willing to accept the productivity concept as long as he could see that unions were not being unfairly treated. Joe Mason was the labor co-chairman of the JALMC from its inception until his death this year. He often said to me that trust was an essential element, and that he would only go along with the joint efforts to improve productivity when he became convinced that the third party facilitators would respect the real needs of workers as well as the sometimes narrow objectives of corporate managers. It seems that the results of declining productivity and the lack of competitiveness are easily apparent; it takes leaders like Joe Mason who understand the problem and can engage in tough bargaining to translate this understanding into a positive process.

The third major implication of the Jamestown experience is that a concentration on improving the quality of working life in individual work places can improve productivity and job satisfaction. I am often asked about the replicability of the Jamestown program. Clearly, it is a program designed to fit local circumstances, including the prevalence of "job shops" and non-assembly line factories. But the principle of QWL seems to me to be applicable in other work places and other areas. I believe that QWL is a process, not a "quick fix" which can be packaged and uniformly applied. Voluntarism is an essential characteristic of the quality of working life movement. It requires commitment and enthusiasm on the part of management on all levels from the first-level foreman to the top executive. You do not get worker participation by executive fiat. Real participation requires an understanding that workers' opinions really will be valued highly, and that a joint approach to redesigning a work place that considers the needs of human beings as well as machines can improve total performance.

We also found in the Jamestown experience that gains sharing and employee ownership can provide real incentives for improving industrial performance. They are not a panacea but are particularly valuable when combined with in-plant, labor-management cooperative examination of the work process.

Finally, there were several advantages of communitywide labor management cooperation that simply could not be realized at the level of the individual enterprise. We found a need for research and skills development that no individual firm could have met with its own resources. Too often government programs concentrated on entry-level jobs in their training program when the real need was for upgrading people who are already employed. Seldom could individual firms realistically assess the impact of technology on the performance of individual workers. It was not always possible for an individual investor who wanted to start an enterprise or take over one that was failing to secure the total trust and cooperation of the company union without the intervention of a third party committed to solving problems on a joint basis.

Many of these forms of labor-management cooperation are still evolving and being tested. Jamestown has a great deal yet to learn in terms of its own development. Two of the largest manufacturing firms have never undertaken in-plant labor management activities, some training programs that have been devised by the labor-management committee have never been able to be implemented, and the committee itself goes through periods of slack as well as positive performance. But it seems fairly certain that the response of one small industrial city to its basic economic problems has developed a form of social innovation necessary to its own survival and applicable to other dimensions of America's economic dilemma.

chapter thirteen

SOCIAL INVENTION AND INNOVATION AS PARTNERS IN TECHNOLOGICAL PROGRESS

Thomas H. Moss

Strategies for strengthening the international competitiveness of U.S. industry are rampant in current public policy discussions. Proposals of this type have been multiplying rapidly since a variety of trade and technological indicators began to show a growing competitive weakness of the U.S. over the last decade. Indices include items as diverse as trade balances,[1] patent statistics,[2] and robot production numbers,[3] among many others. Though not individually decisive, taken together they have given a picture of a rapidly eroding leadership position for this country in industrial technology.

The Carter Administration moved this discussion to the forefront of national policy debate with a well-publicized and extensive "domestic policy review." Literally hundreds of papers and discussions were commissioned in the public and private sector to define an effective national strategy to deal with the problem.[4] This review and follow-up activity from its findings culminated in major program announcements and policy statements in October 1979 and August 1980[5,6] aimed at forming an ambitious "economic revitalization" policy for the country.

Public debate on this issue became a major feature of the 1980 election campaign, and strategies to foster industrial innovation have become a key part of the Reagan Administration's economic program.[7] Congress has also pursued the definition of an innovation and productivity-fostering strategy in its many individual Committees,[8,9] and has groped for forums to integrate individual measures into over-all policies via institutions like the Joint House-Senate Economic Committee[10] and the House Innovation Task Force.[11]

Most of the strategies devised have concentrated very heavily on three elements: capital formation, technological developments, and regulatory policy. These solutions are natural products of the thought of economists, scientists, engineers, and business and political leaders most heavily involved in the discussions. They are undeniably important components of any reasonable strategy.

In contrast, this chapter will stress the role of *social* invention and innovation in enhancing industrial productivity. A growing number of observations emphasize that critical failures in our competitive position can be traced to weakness in the development and utilization of forward-looking social arrangements. Examples include failures to exploit quality control technologies in a wide range of industries, to devise a strategic planning system for the modernization of our steel industry or an effective market forecast system for automobiles, and the lack of reliable mechanisms to resolve labor-management disputes. Social invention and innovation, along with the more familiar technological innovation strategy components, is necessary to overcome such weaknesses if we are to reverse our decline.

In the following, we will look briefly at a few historical instances of the ties between technological innovation and social invention and innovation. It is not difficult to foresee new changes of this type, and we will speculate on a few. More importantly, we will broaden the discussion to two general and fundamental areas of social interaction with technological change. Neither is new, and yet problems of the past make clear that social inventions and innovations are badly needed to improve and modernize them. One of these is the set of arrangements designed to mitigate the disturbing dislocations brought about by technological change. We will stress the crucial importance of these in maintaining a receptive climate for technological innovation and economic development. An even broader need is the related area of societal planning, with its virtues as a participatory and informing process and as a stimulus for progress through goal-setting, but also with its well-known weaknesses of biasing systems toward conservatism and clumsy centralization.

Mechanisms for dislocation mitigation or for broader planning have been frustrating to apply in practice, enough so that a large fraction of a generation of government and private sector administrators have been bred to cynicism concerning the utility or workability of institutional arrangements in these areas. But since technological progress may be impossible without them, improvement in these mechanisms becomes a most important innovation frontier, despite their weaknesses.

Social Invention and Innovation Woven into Modern Industrial Development

Certainly inventions and innovations in the social sphere have been the key to many crucial changes in industrial productivity, from the start of the industrial revolution, to the Hawthorne studies[12] of worker motivation in the 1930s, and continuing through the present. The examples cited here are not meant to be a complete list and are far from static: new elements are being constantly added and old ones sometimes negated as changing conditions lead social inventions and innovations to become the reversal of arrangements that were only recently novel in themselves.

The organization of work into factory production may be the first and perhaps still most profound social invention of the industrial revolution. It is difficult to imagine a more dramatic example of how a social change demanded by technology could profoundly influence the fabric of everyday human life. Steam and water power technologies could not have been of much utility without the social invention of factory organization, and the economic power of machine-assisted production was an irresistible force to stimulate social change over any and all patterns of social conservatism.

Many obvious social inventions have stemmed from the new factory form of work organization. The assembly line and automatic production control are obvious derivatives. So also were the work satisfaction strategies epitomized by the Hawthorne studies and continuing through the many quality control methodologies implemented and discussed in the post-war period in this country and others.

The organizations of labor unions are a similarly related derivative of the original social change. The development of institutions to mediate and negotiate labor-management disputes, including the newer forms such as labor-management councils,[13] followed naturally. Worker participation in management and capital formation, through codetermination arrangements[14] or use of employee pension funds to meet corporate capital needs, are further innovative approaches to the labor-management relationship. All have filled obvious social and institutional needs related to factory and derivative forms of industrial organization; all have created their related derivative demands for complementary, compensating or competitive social inventions.

Social invention and innovation have also characterized the organization of production units, as it has production within those units. Vertical integration of producing subunits is one example, multinational organization of production groups is a more recent one. Sophisticated techniques of market analysis and concerted marketing were necessary social inventions and innovations to support the efficient and geographically integrated manufacturing complexes.

Other more subtle social inventions and innovations have further been linked to technological development in this country. University-industry cooperation was surely a tremendously important social innovation in the agriculture industry,[15] facilitating technological change sufficient to create an industry leadership position for the United States which may be greater and more secure than any since. The Morrill Land Grant Act establishing the Agricultural Universities and State Extension Service[16] must be placed among the most significant of historic social inventions promoting technological change and enhanced economic productivity. As such, it is a classic example of the government-industry-university partnership so often discussed in recent years as an arrangement to achieve gains in manufacturing innovation. Industry-university cooperation has been, in fact, an important social invention in enhancing manufacturing technology and training in cities with strong engineering schools.[17] Though inadvertently weakened by another post-war social innovation, large-sale federal research project support, it may revive again under the twin pressures of federal budget restraint and industrial technology need.

The university has played another socially inventive role in this country's technology base. With the post-war GI ''Bill of Rights'' as the prime stimulus, the education composition of the work force changed dramatically in the post-war years. The percentage of college graduates rose very rapidly from 1 in 20 in the 25 to 29 year age bracket in 1940 to 1 in 4 by 1976.[18] This has led to a ''professionalization'' of the work force in the modern United States, changing the typical character of work dramatically from the factory manufacturing era of the pre-war period,[19] to service, managerial, or information-flow oriented professions of the 1980s.

Many more social inventions and innovations of this type are certainly in developing phases, both driving and being driven by technology. The new technologies of agriculture, ranging from ever higher levels of mechanization to control of products via genetic engineering, may lead to the final industrialization of agriculture, with the demise of the family of "small business" organization of farming. In many other sectors, similarly dramatic impacts may stem from the information and micro-electronics technology revolution. The need for and sociology of meetings, business travel, traditional mail, and many other forms of communication may be dramatically altered. Inexpensive individual access to vast quantities of information and communication channels may once again allow the workplace to be centered in the home. We would then have a new technology reversing the trend started by the similarly technology-stimulated innovation of factory organization of work two hundred years ago.

Each of these and many other cases of coupled social and technological innovation present subtleties in the interplay of social and technological factors that are worthy of study in depth, but are beyond the scope of this essay. Taken collectively, however, they call attention to the need to foster and nurture social invention as a key to providing a climate for introduction of technological innovation and improved productivity. Moreover, looked at with the same favorable bias toward risk-taking which we know pays off in technological innovation, it is clear that the social invention and innovation process should similarly be explicitly acknowledged as an experimental one. Failures or unsuccessful trials in this context should not be considered with dread, but as healthy natural learning experiences, just as they are in technological development. To promote a healthy climate for social innovation we need to replace both the pre-experiment absolutism, and the post-experiment negativism, which so often contrasts discussion of social invention and innovation from the more skeptically experimental technological trials.

Generic Needs for Future Social Inventions and Innovations For Technological Progress

Going beyond the sampling of individual instances of social invention and innovation, improvements in the generic areas of dislocation mitigation and broader societal planning seem essential to pave the way for technological progress. Dislocation mitigation may become especially important in the coming decades as we continue with accelerating change in technology as well as tightening global resource and environmental constraints. Similarly, in planning for the future, we clearly need techniques that help guide our investment of limited resources in new technological developments and yet allow the flexibility toward contingency which is so essential given the frailty of our technological and social foresight. These planning systems should ideally facilitate the allocation of both benefits and liabilities of new technology, and do so in a manner that allows broad societal participation in crucial long-range decisions. Clearly, there is sufficient imperfection in current systems of this kind to warrant extensive social inventiveness in seeking improvements.

The inadequacy in the first of these broad areas, our systems to mitigate the local or individual dislocation effects of technological change, is a fundamental weakness in our ability to introduce technological innovation. The extended political or legal delays to technology implementation have become an all-too familiar phenomenon, at least since the major transportation consolidations of the 1930s.[20] Communities, labor unions, or other groups and individuals have often seen new technologies threaten their vital interest, even in the context of over-all societal benefits. Thus they have not surprisingly acted with all the tools at their disposal to slow and inhibit technological innovation, reinforcing the kind of adversary relationships that have become a hallmark of this country's economic structure: labor versus management, region versus region, sector versus sector, and other conflicts.

It is easy to see how the habit of careless and rapid economic displacement was built up in a country with rapidly expanding frontiers and spectacular economic development. However, this situation becomes very serious in an era where the stress of demographic and geographic factors, as well as energy and materials scarcity, narrow options to buffer the pain of change. The lack of reliable institutions and precedents to provide that buffer makes the threat of change seem locally or sectorally disastrous. The result is that powerful resources are often mobilized against it, despite whatever broad advantages can be forseen. The absence of more effective displacement mitigation mechanisms is a powerful and costly drag on innovative technology.

We are not, of course, entirely without useful arrangements. Consolidation of railroads in the pre- and post-war period,[21] introduction of new shipping technology,[22] and airline deregulation have all caused major labor displacements for which legislated, formalized mechanisms were created for mitigation. Compensation and retraining approaches devised to buffer logging industry workers from job opportunity losses due to park enlargement are another example.[23] However, these organized approaches have usually been restricted to either regulated industries in which the public sector was already heavily involved or disputes in which the government was a party. The precedents and procedures for displacement mitigation are much more undeveloped in the non-regulated manufacturing industry. The primitiveness of our efforts is measured by the weakness of what already exists: rules for layoff notices in many industries typically involving notice as short as a few hours or days.[24]

The tragedy of allowing this kind of limitation on technologically innovative change is that there are many signs that it could be removed with relatively little effort. Labor writings, for instance, often stress open communication as the key criterion for cooperation in technical change.[25] Community planners similarly stress free and timely communication as the key to adjustment to decisions resulting in geographical or sectoral dislocations. Such communication, designed to recognize all parties' legitimate interest and right to be participants in the decisions and conditions of change, are often at least as important as mechanisms of detailed compensation. Other standard measures are more costly: retraining, relocation assistance, or compensation for loss. Nonetheless, taken in total, the dislocation mitigation investment in individual cases is probably more than paid back by the creation of an over-all climate in which innovative change is not feared. Instead, it can be accepted as a healthy evolutionary mechanism in which individuals or localities will be protected from harm.

The fact that the dislocation mitigation investment in a particular case may be substantial, and the benefits for the innovative climate of society as a whole rather diffuse, leads to the difficulty in justifying these investments. It raises the divisive question of needed governmental role even in displacements involving decisions and people entirely in the private sector. This is perhaps where social invention and innovation is most needed. The general difficulties of government or even industry-wide displacement-buffering systems are well known: distorted or inefficient incentives and possibility of uncontrolled cost. Yet the utility is also clear: permitting a climate of innovative change without anxiety, and without placing the full burden of dislocation mitigation on individual innovators as a tax on their innovative capacity. The answer is not likely to be unique or constant in time. Examples exist in which free communication often turned out to be the initial key to opening cooperative problem solving, but solutions have been diverse and imaginative. The common factor has been a willingness to use the same innovative experimental approach to solving the social dislocation problem as may have been used in making the original technical breakthrough.

The discussion of socially inventive mechanisms to allow for displacement mitigation to improve the climate for technological change leads naturally to an even broader social frontier: our systems to plan for the future.

Strategic policy planning has been much maligned in this country, and its application has often had the effect of bringing ridicule to the concept, instead of a learning experience and improvement of the methodology. The inherent faults of policy planning are easy to predict: the approach is normally biased strongly toward conservatism, based on an incremental view of movement to the future. We know historically, however, that social, technological, economic, and physical changes have unfolded with startling suddenness, and our ability to foresee these changes has been very limited. It is often examples of these sudden discontinuities that are used to discredit attempts to institutionalize planning processes, especially at governmental levels. However, the unforeseeable discontinuities have a more serious effect than damage to the image of planning effectiveness. Planning systems that take on a dogmatic and inflexible character actually hinder the adaption of organizations to the threats or opportunities presented by change in external conditions, and thereby become serious inhibitors rather than facilitators of needed technological innovation.

Planning systems have also carried a stigma of association with strongly centralized decision-making mechanisms. Certainly they have historically been used, on both a governmental and industrial level, as devices to insure tight central mangement control over components of large organizations. Aside from insuring an authoritarian cast, under many conditions the centralization itself becomes an inhibitor of innovation. The centralized decision-making apparatus can rarely devise a means to be as effective as decentralized units in quickly and accurately sensing the need or opportunity for innovative change in approaches to problems. This is not to deny the existence of certain counter-examples in which centralization was the key to very rapid adaptive change, but rigidity generally remains an anti-innovative hazard of planning systems that develop[26] with centralized structures.

The existence of these weaknesses should not, however, blind us to the facilitating influence that properly designed planning systems can have on technological innovation. Though a dogmatic planning system can inhibit response to sudden opportunities for technological innovation, a properly adaptive system can function in quite the opposite fashion. Such a planning system allows a framework for seeing a pathway for future development. If its assumptions concerning external circumstances and conditions are properly explicit, it will actually promote adaptation to sudden changes in those conditions by highlighting their significance. As such, the planning system can prevent misallocation of capital or human resources to outdated strategies, and enhance innovation by freeing those resources for the support of promising future-oriented innovation. It can also provide a clear alert to the prospect of social or economic disruption and facilitate measures to minimize them. The planning system can then act to smooth the dislocations that are sometimes such threatening and inhibiting attributes of rapidly developing technology.

Similarly, just as excessive centralization can be a risk of planning systems, in contrast, better design can make such systems vehicles for wide participation in forming and carrying out strategies and for open group or societal communication about goals and assumptions. The process of forming the plan enables participants to express in specific and palpable terms their philosophical beliefs and sense of constraints or opportunities for change. The process can thereby be a means for a group or society to move beyond conflicts over general outlook to workable compromise over realistic choices. The existence of the plan allows an accountability and predictability of leadership actions that can convey to the organization as a whole a much greater sense of identification with over-all goals and strategy than dependence on case-by-case pragmatism.

Institutional improvement for planning need not be viewed as an esoteric issue involving hypothetical new planning boards or agencies. Planning mechanisms, to be built upon and improved, certainly exist at the local government level and in the private economic sector. But even in those areas where planning as an institution seems most primitive or nonexistent (the federal government being a prime example) there is almost always an institutionalized mechanism that can be taken as a least common denominator of planning: the annual budget process.

The budget exercises of both the private and public sector are often the most visible manifestations of planning available, and they share both the weakness and improvement potential of more broadly defined systems. It is not difficult to discover the tendency for symbolism to dominate reality in budget exercises and prose, just as it may in broader or longer range planning operations. It is easy to see that budgeting often becomes a highly centralized and antiparticipatory process, concentrating power and decision-making in a clumsy and nonadaptable fashion. It is also obvious that annual budgeting is usually the epitome of a conservatively-biased planning system. The incremental reasoning so often applied to annual budgeting often leads to simplistic dogma such as "no new starts" in times of budget stress or to the political and bureaucratic difficulty of ending projects even long after the original rationale for their existence has passed.

Principles for Improvement of Planning Mechanisms

Despite the frequently reiterated weaknesses of our planning capabilities and their most frequent crude expression in budgeting processes, it would be a mistake to turn our backs on their importance or sink into unproductive cynicism concerning our ability to make them more effective. Instead, planning mechanisms should command the center of our attention as key frontiers for social invention and innovation. Experiment and improvement in them may be our most important investments in enhancing society's capacity to utilize and adapt to opportunities for technological innovation. There is little doubt that we face, on many levels, a future of danger, resource scarcity, and environmental stress. It is hard to believe that moving into that future with a philosophy of purely random reaction to choices will enable us to survive. This makes improvement of planning mechanisms not merely an interesting sociological exercise, but a necessity for continued existence.

What are the key principles for improving our planning systems, from their expression in budgeting to long-range strategy formation? Perhaps the most generally useful improvement will be to strive to combine a sense of direction with better adaptive feedback mechanisms, which can rapidly indicate the effectiveness of chosen courses. The planning system should be best seen as an expression of Wiener's cybernetic principles:[27] not purposeless or random in initial approach, but rapidly and freely adjusting to changing or more accurately perceived external conditions. The improved feedback system itself will demand further social invention and innovation: better statistics and other forms of information on the state of our society, and perhaps more important, better ways to integrate information so that we sense not a few details but an overall picture of societal vitality.

Building flexibly adaptive feedback into planning mechanisms requires removing the air of dogma which too often accompanies them. Rather than viewing plans as statements of policy to be defended in detail against critics, a concept with greater humility must be sought: the plan as a means of communicating to those both in and outside of an organization an example of how the organization's general principles are applied to a particular set of circumstances. We need to make more explicit that the plan indicates only a set of responses to conditions as they were perceived at the time the plan was formulated. If the conditions change or the perception of them changes, flexibility must be present to allow adjustment of the plan without facing a stigma of inconsistency or lack of foresight.

This leads naturally to a second aspect of planning which sorely needs improvement: the planning process as a means of achieving broad participation in the expression and affirmation of an organization's goals and principles. Even if the flexible feedback-oriented plan is only a snapshot of probable reaction to a set of admittedly transitory circumstances of perceptions, it may be the best means of making clear to all involved the true nature of the group with which they are associated. As such, forming and stating the plan has implications far beyond the issue of whether it is ever used or whether it works well if applied. It is, in this view, a device for internal communication of goals to participants and external communication of intentions to outside observers. The question of the degree of centralization of the planning process then goes far beyond the issue of whether centralization allows for adequate response to external data and efficient implementation of strategy. Much more significantly, the degree of centralization determines the breadth of the base of participants who can be said truly to understand and affirm the organization's guiding principles. The participatory mechanisms devised for forming, articulating, and communicating a plan of action become determining factors in the participatory nature of the organization as a whole.

Devising new mechanisms to allow truly participatory planning processes in an extensive and heterogeneous nation or organization thus becomes the antithesis of the conventional political wisdom of "playing cards close to the vest." Major organizations in government, industry, education, or elsewhere operating with minimally articulated plans convey a message to their own members and the outside world, that, in effect, all but the central leadership will be little more than pawns in a game whose basic strategy is unwritten, with moves predictable only to a select few. Applied to the national government sector, more participatory planning would allow better input of information and ideas from many sectors. At least as importantly, however, it would allow business and other group interests to better predict governmental actions and thus more efficiently allocate their own resources.

It is clear that there are situations where articulating a plan can spoil its chance for success, and other circumstances where the anticipated rewards of political opportunism will seem to dominate any practical or altruistic advantages of participatory planning. However, the prospect of these rewards must be weighed carefully given the significance of planning mechanisms in determining an organization's ability to smoothly handle dislocations brought about by technological change, and in a broader sense in facilitating the internal and external communications of goals. Both are crucial social elements intertwined with the capacity for technological invention and innovation.

Conclusion

Social invention and innovation are not simple concepts to define. Charting strategies to promote wise application is thus compounded in difficulty. Our ability to understand and monitor the human behavioral factors and organizational interactions so important in social invention is primitive compared to our ability to manipulate materials and electronics. It is thus all too easy to concentrate our effort and investment in the easier technological areas and avoid the forbidding frontiers of social invention. Still, the success of societies that have emerged from time to time in world leadership roles can often be traced to an ability to develop and exploit effective new social arrangements which reinforced the value of technological developments. They successfully faced the challenge, in some measure, of harnessing human genius without foundering on human anxiety or other eccentricities. For our own society to ignore this dimension of promoting innovation and productivity, however difficult and frustrating it may appear, would be to risk wasting the potential of many of our bold and exciting technological breakthroughs.

Notes

1. "The Reindustrialization of America," *Business Week,* June 30, 1980, p. 59.
2. M. E. Mogee, Tables of Innovation Indicators, U.S. Library of Congress, Congressional Research Service, December 19, 1979
3. In Robots, "The Cry is Banzai," *Industry Week,* November 27, 1978, p. 85.
4. Domestic Policy Review on Industrial Innovation, initiated May, 1978.
5. Fact Sheet, The President's Industrial Innovation Initiatives, October 31, 1979.
6. Fact Sheet, Economic Program for the Eighties, August 28, 1980.
7. "What Business Wants from Reagan," *Dun's Review,* March, 1961, p. 44-53.
8. Innovation and Productivity Program of the Subcommittee on Science, Research and Technology, *Congressional News Notes on Innovation and Productivity,* House Subcommittee on Science, Research and Technology, June 1, 1981.
9. Activities of the Senate Commerce, Science and Transportation Committee, *Congressional News Notes on Innovation and Productivity,* House Subcommittee on Science, Research and Technology, May 13, 1980.
10. Special Study on Economic Change, Vol. 3, *Studies of the Joint Economic Committee of the Congress of the United States,* Library of Congress, Congressional Research Service, December 29, 1980.
11. *Congressional News Notes on Innovation and Productivity,* Subcommittee on Science, Research and Technology, March 14, 1980.
12. Wendell French, *Personnel Management Process,* (Boston: Houghton Mifflin, 1964), p. 18–19.
13. John J. Popular, "Solution: A Community Labor-Management Committee," *Labor-Management Relation Service Newsletter,* U.S. Conference of Mayors, Washington, D.C., Vol. 10, #11, p. 2 & 3, November, 1979.
14. "Labor's Voice on the Board," *Newsweek,* May 26, 1980, p. 13.
15. Robert E. Evenson, Paul F. Waggoner, and Vernon W. Rutton, "Economic Benefits from Research: An example from Agriculture," *Science* **205,** no. 1101 (1979).
16. Morrell Land Grant College Act of 1862, 12 Stat., 503.
17. Case-Western University and its historic relationship to large industries in Cleveland, Ohio is a good example.
18. Eli Ginzberg, "The Professionalization of the U.S. Labor Force," *Scientific American,* March, 1979.

19. E. Ginzberg, and G. J. Vogta, "The Service Sector of the U.S. Economy," *Scientific American*, March, 1981.
20. Bruce H. Miller, "Providing Assistance to Displaced Workers," *Monthly Labor Review*, May 1979, p. 20.
21. Alice L. Almeety, Worker Protection on Conrail, U.S. Library of Congress, Congressional Research Service Report. No. 81-99E.
22. Bruce H. Miller, "Providing Assistance," p. 18.
23. Ibid; p. 21.
24. Characteristics of Major Collective Bargaining Statistics. *Bureau of Labor Statistics Bulletin* #1957, July 1, 1975, p. 99.
25. Malcolm Peltu, "In Place of Technological Strife," *New Scientist*, March 13, 1980, p. 822.
26. Long Range Planning, House Committee on Science and Technology, 94th Congress, 2nd Session, Serial BB, p. 7–8, 1976.

chapter fourteen

ON USING INTELLECTUAL RESOURCES

Sven B. Lundstedt

For the most part, this book has a practical emphasis. It calls attention in a number of ways to the important idea that technological invention and innovation should be accompanied, followed and even preceded by social invention and innovation. The basic reason for this is obvious enough. It is to help technology to become better adapted to human needs usefully and wisely, and to avoid costly and dangerous side effects whenever possible. The authors have given careful thought to this question. Some have done so from long personal experience with the management and development of technology; others from a research perspective. The point of view that is not as forcefully represented in this book as it might be is that of the social and behavioral sciences, although references are made to them. This is somewhat of an irony, and a limitation of the book, because they have been deeply concerned with social invention and innovation over the past fifty years. In these brief closing remarks we need to call attention to this omission by mentioning their importance.

What are the social and behavioral sciences? We traditionally include in this category social anthropology, economics, psychology, political science, sociology, and at times, history. We are uncertain about history only because, like philosophy, some consider it to be part of the humanities. One recognizes that new disciplines are emerging, but these are still the basic ones.

These disciplines seek to answer questions about man as an individual person, a member of small groups, a member of organizations, a part of larger societies and cultures, and as involved in creating and working within social institutions. Social invention and innovation may be approached at any one or a combination of these levels of analysis. Like the biological sciences, these sciences are important because they focus upon mankind as such, instead of just the physical world and universe. In contrast, the humanities are largely concerned with certain artistic and literary products of mankind or with man as an artistic or philosophical being. The social and behavioral sciences would also favor the view "that the proper study of mankind is man." They are very important intellectual resources that we need to consider in connection with the subject of this book.

William F. Whyte, a sociologist, illustrates how the concept of a social invention can reflect several levels of social analysis:

> I define a social invention as a new and apparently promising strategy designed to solve some persistent and serious human problems. It may take the form of a new organizational structure or a new set of interorganizational relations. It may involve a new set of procedures for shaping human interactions and activities and the relations of humans to the natural and human environment. . . . The potential transferability of a social invention depends upon discovering the theoretical principles underlying its operation and the characteristics of the social and material environment into which it must be fitted in order to solve human problems.[1]

The richness and scope of social and behavioral science contributions are illustrated in an article by political scientist Karl Deutsch and his associates in *Science* magazine in 1971 which refers to social inventions and basic innovations from 1900–1965.[2] Daniel Bell, another sociologist, in an excellent chapter elsewhere on "The Social Sciences since the Second World War," discusses the contributions and highlights the conclusions of this article:

> There are such things as social science achievements and social inventions, which are almost as clearly defined and as operational as technological achievements and inventions.
>
> These achievements have commonly been the result of conscious and systematic research and development efforts by individuals in teams working on particular problems in a small number of interdisciplinary centers.
>
> These achievements have had widespread acceptance or major social effects in surprisingly short times, a range comparable with the median times for wide spread acceptance of major technological inventions.[3]

Some of the social inventions mentioned have a direct, as well as indirect, bearing on technological invention and innovation.

There has been a legitimate concern for many years within American universities and research organizations, and in organizations like the

Social Science Research Council, the Russell Sage Foundation, and the National Science Foundation about the application of social and behavioral research to the problems of society. Even so, the kinds of specific issues discussed in this book have not been recognized as fully as they might be. Evidently, notwithstanding much work already done there is need for even more creative research on the social inventions and innovations which might appropriately fit in with technological invention and innovation to facilitate each and to enhance economic development.

Perhaps it also is well to consider that American social invention and innovation, as contrasted with technological invention and innovation, may occasionally seem out of line with some economic, cultural and political values. One person's social invention may appear to another to be a threat to their well being. In a pluralistic society like the United States this is bound to happen. So there will be times at which social invention and innovation will be resisted and seen as critical of the values, practices, and policies of partisan groups. However, the intellectual freedom to engage in social invention and to take such intellectual risks is very necessary, for without it there will be little progress. There are effective institutional mechanisms in American society for resolving such conflicts of value and social choice if resolution becomes necessary.

While social and behavioral science research has been seen by some as politically or even economicially partisan, for the most part they in fact are not. That they are seen as such is a common error of judgement based on ignorance of both the intent and methods of these sciences. As the accomplishments summarized by Deutsch et al. show, with only a few exceptions the social inventions noted are neither primarily political or particularly ideological in nature. As a rule, the very best of such research follows no particular underlying political or social doctrine, although it is often mistakenly said to do so by critics from the left and right. The social problems associated with modern technological development are of such complexity and importance as to require nothing less than a free spirit of creative and open inquiry. These are the same rules of open scientific inquiry that apply to all other sciences.

Having said this, perhaps we can hope for increased public support of and interest in the part that could be played by the social and behavioral sciences in the solution of the kinds of problems and issues discussed in this book. They are among the most critical social, economic, political, and technological problems of our time. To this end all of our considerable intellectual resources need to be developed. It is our hope that this book will help to identify new directions that may be taken.

Notes

1. William H. Whyte "Whyte aims 1981 program at reorientation of research." *Footnotes* **8**, No. 6, (August, 1980): (Washington, D.C.: American Sociological Association).
2. Karl W. Deutsch, John Platt and Dieter Senghass. "Conditions Favoring Major Advances in Social Sciences." *Science* **171**, N. 3970 (February 5, 1971): 450–59.
3. Daniel Bell "The Social Sciences since the Second World War—Part One." *The Great Ideas Today*. (Chicago: Encyclopedia Britannica, 1979), p. 140.

INDEX

Antiunionism, 106
Apollo program, 3, (see also space program)
Aspen Institute, xi–xii
Atomic Energy Commission, 40
Automobile industry, 5–6, 8, 159–160, 207 208

BASF (Badische Anilin-und Soda-Fabrik), (I. G. Farben), 33–34, 40
Bell, Alexander Graham, 36–37
Bureau of the Budget. See Office of Management & Budget

Carter administration, xiii, 72, 105, 115, 178, 229
Carter, Jimmy, President, xiii, 67
Churchill, Winston, 39
Clean Air Act, (CAA), 153, 154, 159
waiver provisions, 166–167
Clean Water Act, (CWA), 153, 154–155, 159
waiver provisions, 166–167
Comprehensive Employment Training Program, (CETA), 224
Computers, 16, 118, 140–141, 144–145, 187, 210

Department of Housing & Urban Development, 131
Discounted Cash Flow, (DCF), 79–80, 83

Economic decline, 7, 59, 73
Eisenhower administration, 173, 174

Energy,
conservation of, 136,
consumption of, 7–8, 15
increasing costs of, 5, 20
industry regulation, 19
resources, 136–137, 150
savings, 160–163
substitution of labor for, 5
Entrepreneurship, 53–54, 68, 71, 74–75, 86–87
Environmental regulations, 21, 151–169
Environmental values, 120
EPA, (Environmental Protection Agency), 21, 130, 154, 156, 167
bubble concept, 19, 154

Farben, I. G. See BASF
Federal Budgeting Policy, 132, 135
Budget Act of 1974, 133–134, 135
Food technology, 137
Ford administration, 116
"Futurists", 125

Genetics, 139–140, 150
Grants, 47–52, 218–219, 226–227
block, 131
defined, 47
environment, 47
Gray, Elisha, (inventor), 36–37
Gross National Product, (GNP), xiii, 4, 5, 56, 58, 67, 113, 114, 133, 149

Hawthorne studies, 231
Houdry, Eugene Jules (inventor), 37–38

Inertia, 31, 34, 41
Inflation, 69–70, 79, 116, 133
Innovation, 12–13, 45, 54–90
    declining, 4, 67
    defined, v, xiii–xiv, xxi–xxii, 45,
        54–55,
    environmental compliance
        innovation, 157
    fostering, 53, 168–169
    grants economy and, 49–52
    and government, 49, 111–121,
        150–152, 211–212
    and human factors, 93–109,
        207–214
    kinds of, 54
    managerial, 14, 16–18
    political, 18–22
    social, 1–26, 95, 229–240
    stages of, 53
    technological, 1–26, 53–90,
        95–109, 148–169

Jamestown Area Labor-Management
    Committee (JALMC), 217–228
    objectives of, 217
Japan, 190–204
    auto industry, 16
    chemical industry, 85
    economic growth, xiv, 4
    education in, 196–200
    history of, 192–195
    innovation of, 85
    inventiveness of, 190–204
    labor practices, 17
    Quality of Work Life Program, 208
    society, 85–86
    technology & culture of, 200–203
    Westerners' reactions to, 190–191
Johnson administration, 116

Kalmar plant, 209, 211
Kennedy administration, 115–116

Labor, 93, 108
    and innovation, 93–96, 100,
        208–209
    and management, 215, 217–228
Learning curve, 75–76

Man-made disasters, 139

National Aeronautics & Space
        Administration (NASA),
        171–188,
    objectives, 174–175
National Environmental Protection
        Act. See EPA
National Science Foundation, 50, 244
Natural resources, 136–138
Nixon administration, 116, 178
Nuclear power, 120
    and environmental groups, 21
    plants, 39, 137
    Three Mile Island, 209–210

Occupational Safety & Health Act
        (OSHA), 106, 130
Office of Management & Budget
        (Bureau of the Budget), 132,
        133, 144
Ohio State University, xii
OPEC, 48, 136

Patents,
    and genetics, 118
    numbers issued, 114, 229
    policy, 115
Pollution, 74, 153, 161
    abatement, 163–165
    control industry, 159
    (See also Clean Air & Water
        Acts; Environmental
        regulations)
Population growth, 139, 187
Productivity, 5, 98, 207–214
    decline in, 5, 31, 113
    ways to improve, 1–2, 95, 96,
        217, 221, 227

Quality of Work Life (QWL), 208,
    210, 213, 220–221, 224, 227

Radar, invention and development
        of, 35
Reactors,
    breeder, 6
    light water, 6

Reagan administration, xii, 126, 138,
    230
Regulation, 64, 70, 73–74, 114–115,
    126, 134–135, 146
    drafting of, 166
    environmental, 151–169
Rickover, Hyman G., Admiral,
    38–40, 42
Risk & uncertainty, 46
Russell Sage Foundation, 243

Seabed mining, 138
Skill,
    development programs, 223–224
    upgrading, 94, 97
Social and behavioral sciences,
    242–244
Social invention and innovation, 2
    defined, v, xiv–xv, 2, 13–14,
    231–232, 243
    examples of, 9–12, 13–14, 243
Socio technical approach, 10–12, 209
Social Science Research Council, 243
Space program, 6, 117–118, 170–188
    Apollo project, 173, 184, 186
    Project Mercury, 176
    Saturn Project, 173
    Shuttle program, 178, 186
    technology of, 6
    (See also NASA)
Sperry, Elmer, inventor, 34–35, 36,
    42
    and radar, 35

Sperry Gyroscope Co., 34, 36
Sputnik, 171, 173, 174
Sunbelt,
    growth of, 127–129
    moves to, 140

Taxation, 72, 123–124, 128, 154–155
    and pollution controls, 155–156
    Revenue Act of 1969, 72
    Revenue Act of 1978, 72
Technology,
    conservative, 32–43, 64
    and "outsiders," 65–66, 79
    politics and, 123–147
    problem areas, 124–127
    (See also innovation)
"Technology Gap," 3
Toxic Substances Control Act, 155

Unemployment, 89, 94, 98, 102, 103,
    107, 215, 224–225
Unions, 123, 208–209, 211, 213–214,
    219–228, 231
    and innovation, 96–97, 101–102
    (See also Labor)
Urban areas,
    changes & effects on, 127–129
    Model Cities Program, 131
Urban policies, 127, 131

Vietnam War, 132–133

Wye Conference, xv

ABOUT THE AUTHORS

Kenneth E. Boulding
Distinguished Professor of
Economics Emeritus
The University of Colorado at Boulder

Harvey Brooks
Benjamin Peirce Professor of Technology
and Public Policy
Harvard University

The Honorable Clarence J. Brown
Member
U.S. House of Representatives (Ohio)

E. William Colglazier, Jr.
Research Fellow
Center for Science and International Affairs
John F. Kennedy School of Government
Harvard University

Daniel De Simone
President
The Innovation Group
Alexandria, Virginia

Robert D. Hamrin
Economist
Senator Gary Hart and U.S. Senate Budget Committee

Karl G. Harr, Jr.
President
Aerospace Industries Association
Washington, D.C.

Thomas P. Hughes
Professor of the History and Sociology of Science
The University of Pennsylvania

Tetsunori Koizumi
Associate Professor of Economics
The Ohio State University

Ralph Landau
Chairman
The Halcon SD Group Incorporated
New York City

Iris J. Lav
Associate
Ruttenberg, Friedman, Kilgallan, Gutchess & Associates
Washington, D.C.

Virginia C. Lopez
Director, Aerospace Research Center
Aerospace Industries Association
Washington, D.C.

Sven B. Lundstedt
Professor of Public Policy and Administration
The Ohio State University

The Honorable Stan Lundine
Member
U.S. House of Representatives (New York)

Michael Maccoby
Director
Project on Technology, Work and Character
Washington, D.C.

Thomas H. Moss
Staff Director
Subcommittee on Science, Research and Technology
Committee on Science and Technology
U.S. House of Representatives

Frank Press
President
National Academy of Sciences

Stanley Ruttenberg
Partner
Ruttenberg, Friedman, Kilgallan, Gutchess & Associates Incorporated
Washington, D.C.